AF749197

DAM ENGINEERING

Edited by **Hasan Tosun**

Dam Engineering
http://dx.doi.org/10.5772/intechopen.74153
Edited by Hasan Tosun

Contributors

Akira Kobayashi, Akira Murakami, Arash Barjasteh, Aleksandra Ziemińska-Stolarska, Abdikerim Irisbaevich Kurbanaliev, Amanbek Zhainakov, Maralbek Oskonbaev, Zhongzhi Fu, Shengshui Chen, Enyue Ji

© The Editor(s) and the Author(s) 2019

The rights of the editor(s) and the author(s) have been asserted in accordance with the Copyright, Designs and Patents Act 1988. All rights to the book as a whole are reserved by INTECHOPEN LIMITED. The book as a whole (compilation) cannot be reproduced, distributed or used for commercial or non-commercial purposes without INTECHOPEN LIMITED's written permission. Enquiries concerning the use of the book should be directed to INTECHOPEN LIMITED rights and permissions department (permissions@intechopen.com).
Violations are liable to prosecution under the governing Copyright Law.

(cc) BY

Individual chapters of this publication are distributed under the terms of the Creative Commons Attribution 3.0 Unported License which permits commercial use, distribution and reproduction of the individual chapters, provided the original author(s) and source publication are appropriately acknowledged. If so indicated, certain images may not be included under the Creative Commons license. In such cases users will need to obtain permission from the license holder to reproduce the material. More details and guidelines concerning content reuse and adaptation can be found at http://www.intechopen.com/copyright-policy.html.

Notice

Statements and opinions expressed in the chapters are these of the individual contributors and not necessarily those of the editors or publisher. No responsibility is accepted for the accuracy of information contained in the published chapters. The publisher assumes no responsibility for any damage or injury to persons or property arising out of the use of any materials, instructions, methods or ideas contained in the book.

First published in London, United Kingdom, 2019 by IntechOpen
IntechOpen is the global imprint of INTECHOPEN LIMITED, registered in England and Wales, registration number: 11086078, The Shard, 25th floor, 32 London Bridge Street
London, SE19SG – United Kingdom
Printed in Croatia

British Library Cataloguing-in-Publication Data
A catalogue record for this book is available from the British Library

Additional hard copies can be obtained from orders@intechopen.com

Dam Engineering, Edited by Hasan Tosun
p. cm.
Print ISBN 978-1-78985-479-4
Online ISBN 978-1-78985-480-0

We are IntechOpen, the world's leading publisher of Open Access books

Built by scientists, for scientists

4,000+
Open access books available

116,000+
International authors and editors

120M+
Downloads

151
Countries delivered to

Our authors are among the
Top 1%
most cited scientists

12.2%
Contributors from top 500 universities

CLARIVATE ANALYTICS
BOOK CITATION INDEX
INDEXED

WEB OF SCIENCE™

Selection of our books indexed in the Book Citation Index
in Web of Science™ Core Collection (BKCI)

Interested in publishing with us?
Contact book.department@intechopen.com

Numbers displayed above are based on latest data collected.
For more information visit www.intechopen.com

For EU product safety concerns:
IN TECH d.o.o., Prolaz Marije Krucifikse Kozulić 3, 51000 Rijeka, Croatia,
info@intechopen.com or visit our website at intechopen.com.

Meet the editor

Hasan Tosun is a Professor in the Civil Engineering Department of the Faculty of Engineering and Architecture, Osmangazi University, Eskisehir. Dr. Tosun specializes in geotechnics for dam engineering, especially for embankment and rockfill dams and teaches soil mechanics, geotechnics for dam engineering, and engineering geology. Up to 1997, he worked at the General Directorate of State Hydraulic Works, and supervised the geotechnical studies of large dams constructed in Turkey. He has published over 200 technical papers in national and international journals and conference proceedings, and is the author of four books on soil mechanics and geotechnics for dams. He is the president of the Turkish Society on Dam Safety and a member of the National Society of Soil Mechanics and Foundation Engineering in Turkey, the USSD in USA, and CDA in Canada.

Contents

Preface IX

Chapter 1 **Practices in Constructing High Rockfill Dams on Thick Overburden Layers 1**
Zhongzhi Fu, Shengshui Chen and Enyue Ji

Chapter 2 **Seismic Crack Investigation in an Earth Dam by Centrifugal Loading Test 23**
Akira Kobayashi and Akira Murakami

Chapter 3 **Three-Dimensional CFD Simulations of Hydrodynamics for the Lowland Dam Reservoir 39**
Aleksandra Ziemińska-Stolarska

Chapter 4 **Large-Scale Modeling of Dam Break Induced Flows 59**
Amanbek Jainakov, Abdikerim Kurbanaliev and Maralbek Oskonbaev

Chapter 5 **Influence of Geological Structure on Dam Behavior and Case Studies 73**
Arash Barjasteh

Preface

Engineering of dams and their auxiliary structures involves a wide range of processes ranging from planning to operation stages to providing electricity generation, water supply, flood control, and some other services. In other words, it should include not only the construction, but also interdisciplinary studies to ensure that the dam can be used economically for a long time. In addition, dam safety is a very important issue and their construction should not cause problems in terms of public safety.

Dam engineering assists in selecting the appropriate dam location, taking into account the hydrological balance, geotechnical conditions, and economic and environmental impacts that may occur during the operation stage. For this purpose, various computer programs are used to analyze water and ground samples and to investigate potential ground motion problems. In structural and hydraulic design analysis, the behavior of soil and rock fill is evaluated under separate loading conditions. The risk of the dam is assessed against the ground movements associated with the earthquake by means of seismic analysis. It includes material tests with advanced devices for control of soil, rock, and concrete materials to be used during the construction stage, measurement, and observation processes during the construction stage and preparation of an emergency action plan for the operation stage.

This book presents recent research in the different fields of dam engineering and consists of five chapters. Each of these chapters contain valuable study on specified areas of dam engineering. The first chapter involves a study that was performed on constructing high rockfill dams on thick overburden layers, based on Chinese practices. The second chapter presents the results of an investigation performed on seismic cracks in an earth dam by centrifugal loading test, based on Japanese dams. The third chapter deals with the processes of a single-phase three-dimensional CFD model of hydrodynamics in a dam reservoir in Poland. In the fourth chapter, large scale modeling of a dam break is discussed using actual projects in Kyrgyzstan. The fifth chapter focuses on the influence of geological structures on dam behavior when considering large dams in Iran.

I hope that this book provides an opportunity for readers working in the field of dam issues to study recent scientific research and the latest knowledge.

Prof. Dr. Hasan Tosun
Osmangazi University
Eskisehir, Turkey

Chapter 1

Practices in Constructing High Rockfill Dams on Thick Overburden Layers

Zhongzhi Fu, Shengshui Chen and Enyue Ji

Additional information is available at the end of the chapter

http://dx.doi.org/10.5772/intechopen.78547

Abstract

Rockfill dams are very widely constructed all over the world due to their good adaptability to diverse geological and geographical conditions, and their relatively low cost compared to other dam types. However, natural satisfactory sites are increasingly difficult to find in many countries due to past dam development. In some circumstance, building dams over thick overburden layers is unavoidable. In this chapter, Chinese practices in constructing high earth and rockfill dams over thick overburden layers are reviewed. The geological and geotechnical investigation techniques are briefly summarized, and seepage control systems of some selected cases as well as the connection of the impervious systems of both the dams and their foundation layers are described. Commonly used foundation improvement techniques are also presented, followed by simple descriptions of aspects that require further research and development.

Keywords: rockfill dam, overburden layer, seepage control, foundation treatment, in-situ test

1. Introduction

Various types of dams have been increasingly constructed all over the world for irrigation, flood controlling, power generation, environment protection, etc. [1]. Normally, most dams are preferentially built on rock foundations where seepage control is not a very difficult task. However, with the exploitation and exhaust of natural satisfactory dam sites, many new dams have to be constructed on thick overburden layers, as better sites are not available and removal of the existing overburden is technically or economically unfeasible. This adverse situation is often encountered along many hydropower-rich rivers in the southwest and

IntechOpen

© 2018 The Author(s). Licensee IntechOpen. This chapter is distributed under the terms of the Creative Commons Attribution License (http://creativecommons.org/licenses/by/3.0), which permits unrestricted use, distribution, and reproduction in any medium, provided the original work is properly cited.

northwest regions of China [2, 3]. When such thick overburden foundation layers can neither be avoided nor removed, a rockfill dam is often a priority due to its excellent adaptability to such geological conditions. In recent years, more than 50 high rockfill dams, including earth core rockfill dams (ECRDs), asphalt core rockfill dams (ACRDs) and concrete faced rockfill dams (CFRDs), have been constructed in China, as selectively listed in **Table 1**. Challenges can be seen from both the height of these dams and the thickness of the overburden layers.

Technical problems requiring special attention in the design and construction of rockfill dams over thick overburden layers include, but are not limited to, the following aspects:

a. *Shear strength and deformability of load-bearing layers*. The shear strength of underlying foundation layers influences the overall stability of the dam, while the deformability of these layers controls not only the deformation of the dam but also the deflection of the cutoff

No.	Name	Type	Year	H_{max} (m)	T_{max} (m)	Cutoff wall		References
						Width(m)	Height (m)	
1	Shiziping	ECRD	2010	136	102	1.2 × 1	101.8	[4]
2	Xiaolangdi	ECRD	2000	160	80	1.2 × 1	82.0	[5]
3	Pubugou	ECRD	2010	186	75	1.2 × 2	78.0	[6]
4	Changheba	ECRD	/	240	79	1.4 × 1 & 1.2 × 1	50.0	[7]
5	Maoergai	ECRD	2011	147	57	1.4 × 1	52.0	[8]
6	Shuiniujia	ECRD	2006	108	30	1.2 × 1	32.0	[9]
7	Luding	ECRD	2011	84	148	1.0 × 1	110.0	[10]
8	Qiaoqi	ECRD	2006	125.5	72	1.2 × 1	70.5	[11]
9	Xiabandi	ACRD	2009	78	148	1.0 × 1	85.0	[12]
10	Badi	ACRD	/	97	120	1.0 × 1	105.0	[13]
11	Huangjinping	ACRD	2015	85.5	134	1.2 × 1	113.8	[14]
12	Yele	ACRD	2005	124.5	420	(1.0–1.2) × 1	154.5	[15]
13	Xieka	CFRD	2014	108.2	100	1.2 × 1	86.0	[16]
14	Nalan	CFRD	2006	109	24	0.8 × 1	18.0	[17]
15	Miaojiaba	CFRD	2011	111	48	1.2 × 1	41.5	[9]
16	Jiudianxia	CFRD	2008	136.5	56	1.2 × 1	30.0	[18]
17	Aertash	CFRD	/	164.8	94	1.2 × 1	90.0	[19]
18	Chahanwusu	CFRD	2008	110	47	1.2 × 1	41.8	[3]
19	Duonuo	CFRD	2012	112.5	40	1.2 × 1	35.0	[20]
20	Laodukou	CFRD	2009	96.8	30	0.8 × 1	29.6	[21]

Note: ECRD = earth core rockfill dam; ACRD = Asphalt core rockfill dam; CFRD = concrete faced rockfill dam; H_{max} = maximum height of dam; T_{max} = maximum thickness of overburden; '/' means the dam is still under construction and has not been finished.

Table 1. Basic information of typical rockfill dams built on overburden layers in China.

wall, if used. The inhomogeneity of foundation materials can result in differential and incompatible deformation within the dam and may ultimately lead to threatening cracks.

b. *Permeability and erosion resistance of the overburden layers*. One of the most important functions that should be achieved in dam engineering is the ability to control the seepage within the foundation. Designing an impervious system for the dam foundation depends, to a large extent, on the permeability and erosion resistance of the involved strata and the available foundation treatment equipment and techniques.

c. *Liquefaction potential of the underlying fine layers*. Earthquake is one of the most disastrous natural events that dams are expected to experience. Cyclic shearing by earthquakes can cause excessive pore water pressure to build up in fully saturated sandy soils, leading to a decrease, or even loss, of their shear strength. As a result, uncontrollable deformation can occur in both the foundation and the dam itself and may result in the worst-case scenario of a dam breach.

d. *Connection techniques for the impervious systems of the dam and its foundation*. An effective impervious system means not only successful control of the seepage through the dam and its foundation but also satisfactory performance of the connection points between different impervious components. These points are usually places where parts with different rigidity levels meet and joint, and are therefore vulnerable to cracks and concentrated leakage.

The main challenge in constructing a rockfill dam on thick overburden layers is the design and successful construction of an impervious system for the foundation, accounting for the distribution of the underlying soil and rock layers as well as their physical and engineering properties. In this chapter, the authors review several high rockfill dams built on thick overburden layers in China in order to provide a reference for similar cases that might be encountered in the future. The chapter starts with general descriptions of some frequently implemented geological and geotechnical investigation techniques. Next, seepage control techniques used in some selected cases are introduced. Attention is also paid to the connection techniques for impervious systems used in different kinds of rockfill dams and to the widely adopted foundation reinforcement measures in engineering practice. Directions that deserve further research and development are presented.

2. Geological and geotechnical investigations

Thick overburden layers generally refer to quaternary materials deposited over river beds, including boulders, cobble, gravel, sand, silt and clay constituents. Mixtures of these complex overburden materials are often much more compressible and permeable than an intact rock foundation. Adequate geological and geotechnical investigations on the distribution, thickness and other relevant properties of the soil strata are necessary for the design of impermeable systems and for the preparation of required foundation treatments during dam planning stages. In particular, weak layers, such as sand lenses, soft clays and collapsible loess, should be revealed in these investigations and then properly treated to eliminate safety risks to both the foundation itself and to the overlying dam.

2.1. Geological investigation

Core drilling is the most useful subsurface exploration method for investigating the location, extent and constituent makeup of soil and rock strata at a potential dam site. Nonetheless, core drilling becomes increasingly difficult through overburden layers thicker than 40–50 m because [22]: (1) the existence of unpredictable super-large rock particles; (2) frequent borehole collapse; and (3) uncontrollable loss of drilling fluid. The mud or water used in ordinary drilling-with-casing operations can also make the analysis of core grading difficult or inaccurate due to washing away of fine particles. Some special core drilling techniques were therefore used to get high quality cores in the geological investigations at Yele, Aertash, Xiabandi and other dam sites. They include: (1) double-tube swivel type diamond drilling with a proper rpm (revolution per minute) and pressure and (2) special vegetable gum and powder drilling fluid circulated under proper flow rates to protect the bit, the borehole and the core. Until now, the deepest overburden core drilling conducted in China is at Yele ACRD, where the overburden thickness reaches 420 m [15].

Geophysical exploration methods, such as electrical and electromagnetic methods, seismic procedures, gravity techniques, magnetic methods, and so on, are now increasingly used in dam engineering. These techniques are mostly used to locate the interface between overburden and bedrock and to detect weak layers. Geophysical techniques generally does not directly measure the parameters desired for designing purpose. The vast majority of objectives is inferred from the known geologic data and measured geophysical contrast [23]. That is to say, an inverse solution is sought usually in geophysical exploration, and in most cases, it is the most likely but not necessarily the unique conclusion. Assumptions used in interpreting geophysical contrasts, such as the distinct subsurface boundaries, the homogeneity of materials and the isotropy of material properties, are also, in many cases, at variance with the reality, which may lead to inaccurate and misleading conclusions. Therefore, geophysical methods are almost always used in combination with irreplaceable core drilling. Thereby, results obtained by different methods can be verified mutually and a most reliable judgment can be made.

2.2. Geotechnical tests and interpretation

While geological explorations give overall information on the overburden layers, geotechnical tests and their interpretation yield more relevant parameters for designing. However, systematic laboratory experiments with overburden materials are usually unrealistic due to difficulties in obtaining high-quality undisturbed samples. Although some techniques do exist for sampling (e.g., in-situ freezing [24]), they are generally expensive and only applicable to shallow layers. Therefore, measurement of engineering properties of overburden materials relies more upon in-situ tests as exemplified as follows:

a. *Heavy and super-heavy dynamic penetration tests*. For layers with high relative densities, heavy or super-heavy dynamic penetration tests are usually used, in which a cone-tipped probe is driven into the ground by a 63.5 or 120 kg weight dropped freely from a height of 76 or 100 cm. The number of blows needed to drive the probe into the tested layer for 10 cm is registered and an average penetration per blow is calculated. The data gathered can then be used to estimate the density states, bearing capacities, and moduli of the tested layers. This method has been used in geotechnical investigation of almost all dam sites [25].

b. *Plate load tests*. Plate load tests are performed by loading a steel plate at a particular depth and recording the settlement corresponding to each load increment. The load is gradually increased until the plate starts to sink at a rapid rate. The total load on the plate at this stage, divided by its area, gives the value of the ultimate bearing capacity of the tested soil. Assuming an isotropic elastic behavior of the tested soil, the elastic modulus can also be evaluated. In the under-construction Aertash CFRD, plate load tests were performed with a plate of 1.5 m in diameter, and with the maximum reaction force of about 1000 tons [26].

c. *In-situ shear tests*. Large-scale in-situ shear tests are widely used in field investigations. A shear box of a specific size is compressed into the overburden and then it is pulled, after applying a designed vertical load upon the enclosed soil, by a jack using a high-strength chain. The applied horizontal force and the displacement of the shear box are recorded, based on which the in-situ shear strength of the tested soil is determined [27]. A special advantage of this method is that it can measure the strength of coarse granular materials under extremely low normal stresses, which is otherwise not so easy in triaxial compression experiments.

d. *Pressuremeter tests*. Pressuremeter tests are performed in-situ by placing a cylindrical probe in the ground and then expanding the cylinder to pressurize the soil horizontally. The radial pressure on the soil and the relative increase in cavity radius are measured, from which the in-situ stress strain curve of the soil is derived. This technique is extremely attractive in testing overburden layers because the loading direction is identical to the hydrostatic pressure upon a cutoff wall if it is to be installed. Abundant information can be obtained from this type of tests, such as the in-situ horizontal stress, the pressuremeter modulus, the limit pressure, etc. This technique has been used in many dam sites [22, 28], and the depth has reached a magnitude of 100 m successfully.

e. *Wave velocity tests*. The most widely used wave velocity tests include the down-hole test, the suspension logging test and the cross-hole test. The first two methods require only one borehole and evaluate the wave velocities vertically along the borehole wall. The third test requires at least two boreholes and measures the wave velocities within a horizontal plane. Boreholes for down-hole and cross-hole tests should be carefully cased and grouted to ensure good seismic coupling between the geophones and the surrounding soils. Suspension logging, on the other hand, preferably uses uncased holes. All three methods have been applied to the investigation of overburden layers for foundations in many important projects [16, 25]. The velocity results obtained, especially the shear wave velocities, can be used to evaluate the density states, the elastic moduli and the liquefaction potential of the tested layers.

f. *Permeability tests*. The permeability coefficients of overburden layers, which are needed to design the underground impervious systems, are determined by various permeability tests. Methods may be selected based on the location of underground water table, the enrichment of underground water, and the hydraulic conductivity of the concerned layer. In principle, pumping tests or water injection tests are conducted to evaluate the permeability of highly permeable overburden layers, while pump-in tests are used for relatively less permeable bedrock layers [25]. Permeability tests are indispensable for almost all dam projects.

Most of in-situ geotechnical investigation techniques listed above require high-quality pre-drilled boreholes. Unfortunately, this becomes increasingly difficult when the thickness of the overburden at the potential site exceeds 50 m. Uncertainty exists in all foundation conditions, and therefore designing and constructing an underground impervious system within a thick overburden is a very challenging task. Adequate geological and geotechnical investigations are undoubtedly the only way to improve design confidence in these systems. It is also important for designing engineers to fully assess the reliability of investigation results, including factors such as the size effect in plate load tests, the field draining condition in pressure meter tests, the influence of underground water on compressive wave velocities, the influence of drilling fluid layer adhering to the borehole wall on measured permeability coefficients, and the possible anisotropy of engineering properties.

3. Seepage control techniques

The two main goals in designing seepage control facilities include controlling the hydraulic gradient within overburden layers to ensure the seepage stability of foundation materials and reducing the seepage loss of reservoir water. For overburden layers where excavation and removal are feasible, deposits right beneath the impervious system of the dam body (e.g., clay or asphalt core and toe plinth) can be removed so that the seepage barrier can sit on a firm rock foundation. In cases where deep excavation is impossible, a horizontal, vertical or combined seepage control measure must be employed to meet the above goals. In the design specification for rolled earth and rockfill dams, a vertical seepage barrier that cuts through the overburden layers is recommended over an upstream horizontal measure. This can be evidenced from the cases listed in **Table 1**, in which all dams use at least one cutoff wall. In this section, seepage control techniques used in some of these example dams are reviewed.

3.1. Earth core rockfill dams

3.1.1. The Xiaolangdi ECRD

The Xiaolangdi ECRD was constructed on the well-known, sediment-laden Yellow River. The thickness of the underlying overburden is approximately 80 m, and it is composed of intricate sand and gravel layers. The dam uses an inclined core wall (low-plasticity loam) as the main anti-seepage barrier, as shown in **Figure 1**. A vertical concrete cutoff wall (1.2 m) was built within the overburden to control the underground seepage. The top of the cutoff wall was embedded into the core wall for 12 m, while its bottom end penetrates the rock surface for at least 1 m. The inclined core was extended using low permeable clayey soils along the top surface of the cofferdam on the upstream side, forming a horizontal blanket that is useful in lengthen the seepage path. The cutoff wall under the cofferdam was elongated into this

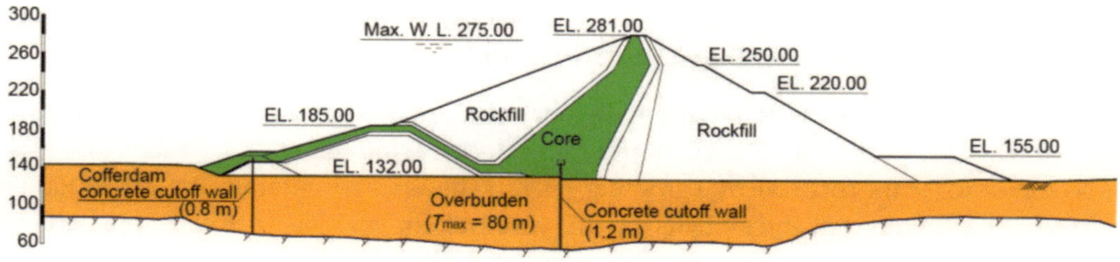

Figure 1. The maximum cross section of the Xiaolangdi ECRD.

blanket. It was assumed that the upstream blanket would connect naturally with sand sediments during long-term operation once the reservoir was impounded.

During the design-phase for the Xiaolangdi ECRD, China had no experience in building such high rockfill dams on 80-m overburden layers, making this project a particularly difficult challenge. A number of alternative design proposals were also considered, including the complete removal of the overburden under the core wall and the use a horizontal impervious blanket without the permanent cutoff wall. Lessons learned from previous cases and, more importantly, technological advances in cutoff wall construction resulted in the final chosen design. The thickness of the cutoff wall was determined based on the allowable hydraulic gradient of concrete materials, the available equipment and stress–strain and seepage analyses results. Conventional concrete with a 28-d strength of 35 MPa was used for the main cutoff wall, while plastic concrete and high-pressure rotary jet grouting were used to construct the temporary cofferdam cutoff wall.

3.1.2. The Changheba ECRD

The Changheba ECRD is currently one of the highest rockfill dams under construction in China (**Table 1**). It sits on a thick, three-layer overburden. All three layers, $fglQ_3$, alQ_4^1 and alQ_4^2 (shown in **Figure 2**) consist mainly of coarse gravel materials and therefore have relatively high deformation moduli and bearing capacity, but also exhibit high permeability. Local liquefiable sand layers are, however, also distributed widely within the alQ_4^1 layer. The dam is located in a high earthquake intensity region, where the peak acceleration for an exceedance probability of 0.02 in 100 years is 3.59 m/s^2. Sand liquefaction under earthquake condition is therefore a potential problem for this dam. The existence of these sand layers may also cause uneven deformation of the dam. To avoid these adverse risks, sand layers beneath

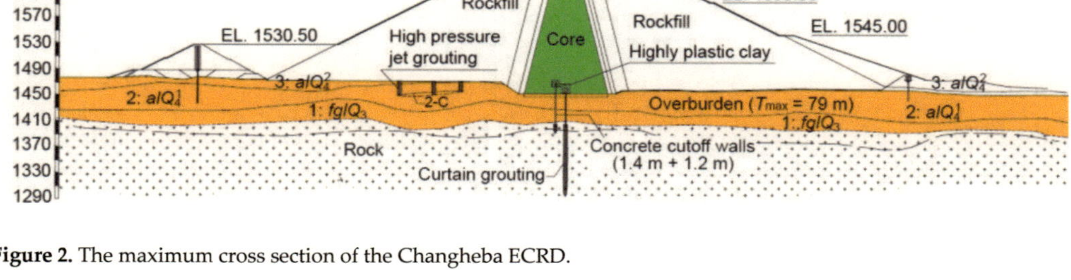

Figure 2. The maximum cross section of the Changheba ECRD.

both the core wall and the filter layer were removed completely. The maximum thickness of the retained overburden under the core wall is about 53 m.

Two concrete cutoff walls were poured with a net distance of 14 m. Both walls penetrate into the bedrock for at least 1.5 m. The main cutoff wall (1.4 m) is located within the dam axis plane and is connected to the core wall by a grouting gallery. The auxiliary cutoff wall (1.2 m) is located upstream of the main wall and embeds into the core wall for 9 m. The core wall is constructed with gravelly soils, where the maximum core material diameter allowed is 150 mm. The percentage of particles finer than 5 mm (P_5) ranges from 30–50%. Another two strict requirements for the core materials are $P_{0.075} \geq 15\%$, and $P_{0.005} \geq 8\%$. Curtain grouting was conducted through the preset pipes within the cutoff walls. In particular, curtain grouting under the main cutoff wall was extended to the level 5 m below the relatively impermeable layer ($q < 3$ Lu.)

3.1.3. The Luding ECRD

The 84-m Luding ECRD was built on an overburden with a maximum thickness of 148 m. It is among the deepest overburden layers used as foundations of a rockfill dam in China. The complex soil and rock strata is shown in **Figure 3**. Four main layers can be observed: the $fglQ_3$ layer, the $prgl + alQ_3$ layer, the $al + plQ_4$ layer, and the alQ_4^2 layer. Basic properties of these layers are listed in **Table 2**. The third sub-layer of the $prgl + alQ_3$ layer consists mainly of fine and silty sands, and therefore has a relatively low deformation modulus and a low bearing capacity. Sand lenses also exist in the second sub-layer of the $prgl + alQ_3$ layer and the first sub-layer of the $al + plQ_4$ layer.

The dam uses a clay core as the anti-seepage barrier stabilized by the rockfill shoulders. The maximum diameter allowed for the core materials is 100 mm. Other restrictions imposed on the core materials are $P_5 \geq 90\%$, $P_{0.075} \geq 60\%$, and $P_{0.005} \geq 15\%$. Repeated compaction near the optimum water content (±2%) produces a barrier with a coefficient of permeability less than 5×10^{-7} cm/s. A vertical concrete wall (1.0 m) was designed to cut off the foundation seepage water. The cutoff wall penetrates into the bedrock at both the left and right abutments of the dam. However, the overburden near the center of the canyon is so thick (148 m) that the current technology limits the capacity for constructing such a high underground wall. Therefore,

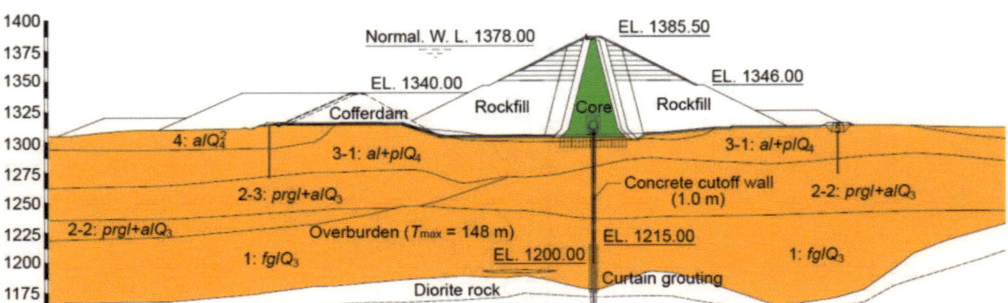

Figure 3. The maximum cross section of the Luding ECRD.

http://dx.doi.org/10.5772/intechopen.78547

Layer	Density		Modulus and bearing capacity		Shear strength		Permeability	
	ρ (g/cm³)	ρ_d (g/cm³)	E_0 (MPa)	R (MPa)	φ (°)	c (MPa)	k (cm/s)	J_c
1: $fglQ_3$	2.20–2.30	2.05–2.15	55–65	0.55–0.65	30–32	0	2–4 × 10^{-2}	0.12–0.15
2-2: $prgl + alQ_3$	2.05–2.15	2.00–2.05	40–50	0.35–0.45	26–28	0	1–5 × 10^{-3}	0.20–0.25
2-3: $prgl + alQ_3$	1.60–1.70	1.40–1.60	18–22	0.12–0.16	15–18	0	1–10 × 10^{-3}	0.25–0.36
3-1: $al + plQ_4$	2.10–2.20	2.05–2.10	45–55	0.40–0.50	29–31	0	5–10 × 10^{-3}	0.15–0.18
4: alQ_4^2	2.15–2.25	2.00–2.10	50–60	0.50–0.55	28–30	0	1–10 × 10^{-2}	0.10–0.12

Note: ρ = natural density; ρ_d = dry density; E_0 = deformation modulus; R = allowable bearing capacity; φ = friction angle; c = cohesion; k = coefficient of permeability; J_c = allowable hydraulic gradient.

Table 2. Basic properties of the overburden layers in Luding ECRD.

a suspended cutoff wall was designed in the river center with the bottom end located at an elevation of 1200 m within the $fglQ_3$ layer. The cutoff wall was connected to the clay core by a grouting gallery. The maximum height of the wall is 110 m, and the underlying unsealed overburden has a thickness of 40–50 m. Two rows of grouting pipes (φ 114 mm) were preset in the cutoff wall for grouting the bedrock, and two additional rows outside the wall for grouting the unsealed overburden. Both curtains extend into the bedrock, that is, the rock curtain reaches the level where $q < 5$ Lu., and the overburden curtains penetrate the rock for at least 2 m.

3.2. Asphalt core rockfill dams

3.2.1. The Yele ACRD

When high-quality clayey soils are difficult to obtain to construct an ECRD, an ACRD is an appropriate alternative. Asphalt is a highly plastic and impermeable material and has a good

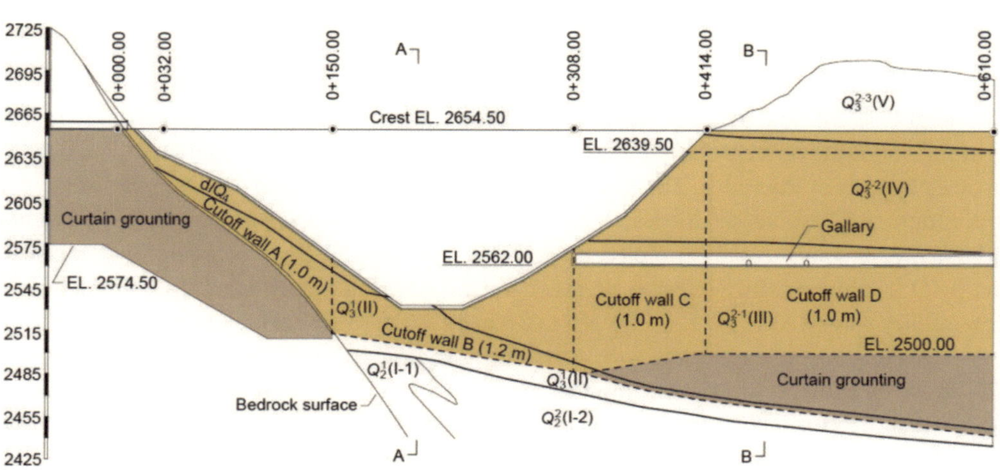

Figure 4. The segments of seepage control barriers in the Yele ACRD.

adaptability to uneven deformation. The Yele ACRD sits on a thick overburden as shown in **Figure 4**, with an extremely thick overburden at the right abutment. There are five main layers under the dam, as divided by the solid curves in **Figure 4**. The first (Q_2^1 & Q_2^2), third (Q_3^{2-1}), fourth layers (Q_3^{2-2}) are mainly composed of weakly cemented gravel materials, while the second layer (Q_3^1) is composed of a mixture of gravel and hard clay. The relatively high fifth layer (Q_3^{2-3}) is mainly composed of silty loam. The second layer (Q_3^1) forms a relatively impermeable barrier in the foundation, the permeability coefficient of which is less than 2.2×10^{-5} cm/s and the allowable hydraulic gradient reaches 10.4. These features were fully used in designing the foundation impervious facility.

The seepage control measures for this dam are divided, from the left bank to the right, into a number of different segments as described below. Curtain grouting was conducted within the gallery in the left river bank (0–150.00–0 + 007.275) to an elevation of 2574.5 m, with a maximum depth of 80 m. From 0 + 007.275 to 0 + 150.00, a concrete cutoff wall (1.0 m) was built into the bedrock for 1.0–2.0 m and curtain grouting was conducted into the weakly weathered rock. The maximum height of the cutoff wall in this segment is 53 m. The third segment starts from 0 + 150.00 until 0 + 308.00, and has a suspended cutoff wall (1.2 m), with its bottom end penetrating the second layer (Q_3^1) for at least 5 m. The height of the cutoff wall in this section ranges from 25 m to 74 m, and curtain grouting was not conducted. From 0 + 308.00 to 0 + 414.00, two layers of concrete cutoff wall were constructed separately. The lower cutoff wall (1.0 m) was cast within the gallery, while the upper wall was constructed from the slope surface. Curtain grouting was conducted into the second layer (Q_3^1) for at least 5 m. The fourth segment (0 + 414.00–0 + 610.00) uses a similar combination of two layers of cutoff wall (1.0 m) and a curtain grouting. The lower cutoff wall was cast to an elevation of 2500 m in the gallery, beneath which a curtain grouting embedding the second layer for at least 5 m was used to cut off the seepage water. The maximum depth of the curtain grouting in this fifth segment is about 120 m. Reinforced concrete was used for the top of the cutoff wall at an elevation of 2639.50–2654.50.0 m.

3.2.2. The Xiabandi ACRD

The Xiabandi ACRD was constructed mainly with gravel materials collected from the riverbed. The thickness of the foundation overburden reaches 148 m, which is almost twice the dam height (78 m). The distribution of the deposited layers is shown in **Figure 5**, where three main influential layers can be seen. The lowest layer ($fglQ_3^1$) mainly contains glacial gravel particles 2–8 cm in diameter. The thickest layer (glQ_3) mainly consists of coarser grains such as boulders and rubble, and it has local bridged structures distributed widely throughout and has a very complicated lithology. Enclosed within the glQ_3 layer is an almond thick sand lens ($fglQ_3^2$) which mainly consists of medium and fine sand, silty loam and silty sand. No high-quality clayey soils are found within 60 km of the dam site, and cement, steel and other necessary construction materials would also have to be imported from places even far away (320 km). The transportation condition to the dam site is rather severe at the time of designing. Traffic interruption is often caused by heavy snows in winter while in summer the flood originated from melting ice and snow often results in debris flow accidents. Because of these natural conditions, using too much steel and cement should be avoided. The dam site, on the other hand, is rich in good aggregate for asphalt concrete. Therefore, an asphalt core is used as the impervious system of the dam.

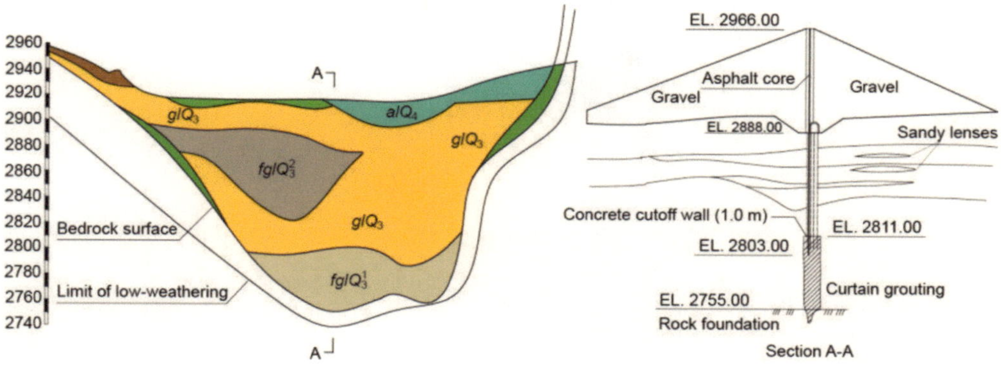

Figure 5. The overburden and seepage control barrier in the Xiabandi ACRD.

A concrete cutoff wall was constructed, with the bottom inserted into the bedrock within shallow bank slopes. At the deepest locations in the center of canyon, concrete was poured from an elevation of 2803 m to an elevation of 2888 m at an ascending speed of 2.0–7.5 m/h, forming an 85-m high-suspended concrete cutoff wall (1.0 m). Four rows of curtain grouting were constructed to extend the impermeable system into the bedrock, including a row of curtain grouting upstream of the cutoff wall and two rows downstream. The middle curtain grouting was performed through the pipes preset in the cutoff wall. The main (inner) curtain grouting penetrates the bedrock for 10 m, and the outer three rows for at least 5 m. The permeability restriction on the curtain grouting is $q < 5$ Lu or $k < 10^{-4}$ cm/s.

3.3. Concrete faced rockfill dams

3.3.1. The Aertash CFRD

The Aertash CFRD, currently under construction, is the highest dam of its type filled upon thick overburden layers. The alluvial foundation materials can be broadly divided into two layers. The upper layer (alQ_4) mainly consists of gravel materials inlayed by boulders, the thickness of which ranges from 4.7 to 17.0 m. The lower layer (alQ_2) is constituted mainly by weakly cemented gravel materials. The total thickness of the overburden layers reaches 94 m, as shown in **Figure 6**. The basic properties of both layers are given in **Table 3**. In general, both gravel layers are in medium dense states and have relatively high strength and deformation moduli. The permeability, however, is also very high and the discontinuous grading makes them vulnerable to seepage failure.

Reinforced concrete face slabs are used to retain the reservoir water and a deep concrete wall (1.2 m) penetrating the rock foundation is designed to cut off the underground seepage. The thickness (t) of the concrete face is $t = 0.4 + 0.0035H$, where H is the depth measured from the top of the face slabs. The concrete face slabs are connected to the concrete cutoff wall by a toe plinth and two horizontal linking slabs. The maximum height of the cutoff wall is 90 m, with the top 10 m reinforced by steel rebar. Curtain grouting is conducted under the cutoff wall into the bedrock to a level where $q < 5$ Lu. The depth of curtain grouting ranges from 17 to 69 m.

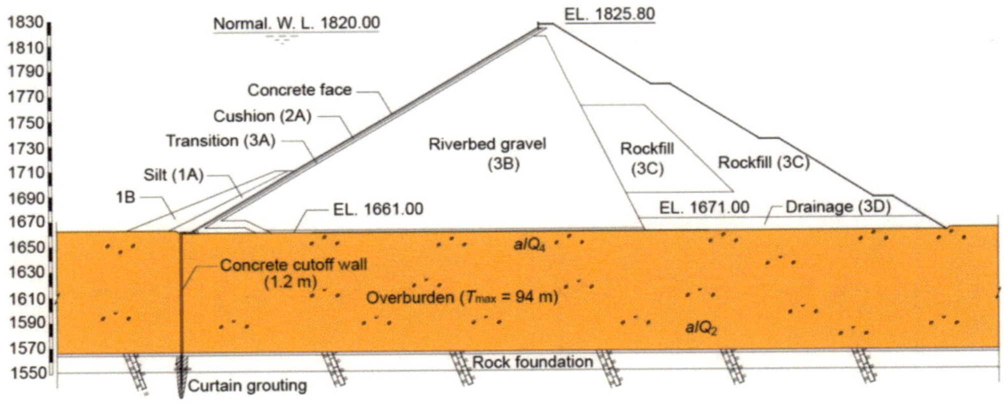

Figure 6. The maximum cross section of the Aertash CFRD.

Layer	Density		Modulus and bearing capacity		Shear strength		Permeability	
	D_r	ρ_d (g/cm³)	E_0 (MPa)	R (MPa)	φ (°)	c (MPa)	k (cm/s)	J_c
alQ_4	0.80–0.85	2.23–2.23	40–50	0.60–0.70	37.0–38.0	0	0.29	0.10–0.15
alQ_2	0.83–0.85	2.18–2.20	45–55	0.65–0.80	37.5–38.5	0	5.00	0.12–0.15

Note: D_r = relative density.

Table 3. Basic properties of the overburden layers in Aertash CFRD.

For concrete faced rockfill dams, it is possible to construct the dam first and then continue with the construction of the cutoff wall, or vice versa. Finite element analyses can be used to optimize the construction sequences. In the current case, the concrete cutoff wall is planned to be built after the dam is filled to a certain elevation. The linking slabs will be cast before reservoir impounding. Connecting the top of the concrete cutoff wall to the linking slabs will also be finalized before impounding.

3.3.2. The Chahanwusu CFRD

The Chahanwusu CFRD is another high dam (110 m) built mainly with gravel materials, as shown in **Figure 7**. The dam sits on sand and gravel overburden layers with a maximum thickness of about 47 m. Three layers can be observed in **Figure 7**: the upper sand and gravel layer with an average thickness of 19.2 m; the medium-coarse sand layer with an average thickness of 5.9 m; and the lower sand and gravel layer with an average thickness of 11.2 m. Both of the sand and gravel layers have similar engineering properties. The average relative density is 0.85 and the average coefficient of permeability is 6.68×10^{-2} cm/s. The middle sand layer has an average relative density of 0.92 and a permeability coefficient of 4.27×10^{-2} cm/s. Therefore, all foundation layers are in relatively dense states. The dam is located within a region of high earthquake intensity, with a design horizontal acceleration of 2.31 m/s². However, liquefaction within the medium-coarse sand layer is considered impossible.

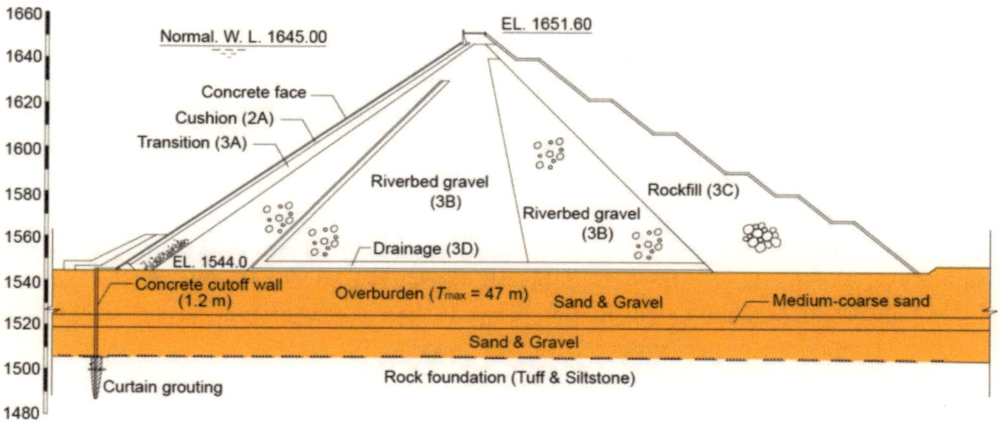

Figure 7. The maximum cross section of the Chahanwusu CFRD.

The dam uses upstream concrete face slabs as the seepage barrier, the thickness of which is determined by $t = 0.3 + 0.003H$. The toe plinths on the left and right bank slopes sit on bedrock, with both consolidation and curtain grouting performed underneath. The toe plinth built on the riverbed is located directly on the gravel layer, removing only the surficial loose deposits (1–2 m). Dynamic compaction was, however, performed to enhance the relative density and modulus of the materials beneath the toe plinth. A concrete wall (1.2 m) inserting the bedrock was constructed to cut off the foundation seepage. The cutoff wall was also connected by two horizontal linking slabs and the toe plinth to the upstream concrete face slabs, forming a closed impervious system. Curtain grouting was also performed under the cutoff wall into the bedrock until the designed level was achieved.

3.4. General remarks

There are other types of rockfill dams and sluices built on overburden layers. Reviewed above are three main kinds of rockfill dams used in water conservancy and hydropower engineering. All the dams in operation reviewed above function well without abnormal performance and major accidents. It could be remarked, in a general sense, that constructing high rockfill dams upon thick overburden layers is technically feasible. Using one or two vertical cutoff wall(s) embedding into the bedrock layer is an effective measure to control the underground seepage. In the case that the underlying overburden layers are extremely thick, a suspended cutoff wall extended by several rows of curtain grouting seems to be a feasible and effective choice.

4. Connection techniques

A reliable connection between the seepage control components within a rockfill dam and its overburden foundation is a prerequisite for a successful impervious system. Connection zones are weak places that require special design considerations. In this section, connection techniques used in different types of rockfill dams are briefly introduced.

4.1. Connection to clay core

For earth core rockfill dams, two basic design schemes can be used to connect the cutoff wall with the clay core. The simplest one is to insert the top of the cutoff wall directly into the clay core for a specified depth. The depth can be determined by the allowable hydraulic gradient along the interface between the cutoff wall and the surrounding soil. Inadequate inserting depth may result in seepage erosion along the contacting path. To avoid shear failure and cracks in the clay core adjacent to the inserting points, a zone of highly plastic clay is used to encapsulate the top of the cutoff wall, as shown in **Figure 8(a)**. Highly plastic clay is more deformable than the clay core, and it can absorb incompatible deformation between the cutoff wall and the core wall without sacrificing its impermeable performance, even under large shear strains.

The second connection method is to use a concrete gallery on the top of the cutoff wall, as shown in **Figure 8(b)** and **(c)**. Using a gallery near the base of the dam has several advantages. Curtain grouting can be performed within the gallery at the same time of dam filling, which may considerably shorten the construction time. Second, the gallery provides a possibility to enforce the foundation impervious system in the case that it does not function well. Without a gallery, repairing the underground seepage control component will be extremely difficult, if not impossible. The gallery can also be used to monitor the performance of the dam and it also provides a path to connect the left and right bank slopes. In the Changheba and Pubugou ECRDs, two concrete cutoff walls are used, one inserting into the clay core and the other connected with a gallery enclosed by highly plastic clay zones.

Careful designing, however, should be exercised when using a gallery. On the one hand, no structural joints are used in general for the riverbed monolith, and the gallery is vulnerable to cracks because of uneven settlement of the overburden layers. The gallery is usually extended into the rock banks, and the connection places of riverbed and rock segments often suffer large shear deformation and damage of water stops. On the other hand, connection of the concrete cutoff wall and the floor of the gallery also require special design considerations. In current practice in China, a rigid connection is mostly wide used, where the top of the concrete wall is reinforced and cast together with the floor of the gallery using an inverted trapezoidal transition cap, as shown in **Figure 8(b)** and **(c)**. A rigid connection may result in

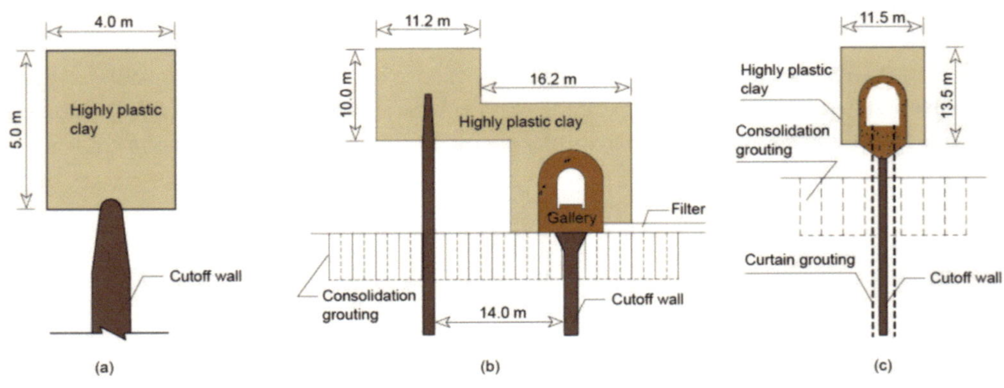

Figure 8. Connection techniques for ECRD. (a) Xiaolangdi ECRD. (b) Changheba ECRD. (c) Luding ECRD.

high compressive stresses within the cutoff wall, but it removes the need for a complicated water stop structure between the cutoff wall and the gallery floor.

4.2. Connection to asphalt core

An asphalt core is a thin plate structure similar to a concrete cutoff wall. A connection between the two structures is often accomplished using a concrete base built on the top of the cutoff wall, as shown in **Figure 9(a)**. The location of the concerned section can be seen in **Figure 4**. An inverted trapezoidal cap is used to accommodate the enlarged foot of the asphalt core wall, with mastic asphalt used to ensure the cementation between the cap and the core wall. The top surface of the base is usually curved slightly downwards to ensure that the asphalt core does not spread. A gallery can also be incorporated into the concrete base for inspection, grouting and communication. Asphalt core has currently not been used in rockfill dams higher than 150 m in China, and the connection with concrete cutoff wall is usually simpler than that in ECRDs as described above.

Another distinct feature of the reviewed Yele ACRD is the use of two layers of concrete cutoff walls that are connected by a construction gallery in the right bank, as shown in **Figure 9(b)**. The upper cutoff wall was constructed on the ground, while the lower one in the gallery (6.0 × 6.5 m). Curtain grouting was also finished in this gallery. To connect the upper cutoff wall with the top of the gallery, joint curtain was constructed using a non-circulation descending stage grouting. Both cement and chemical slurries were pumped into the jointing soils under a maximum pressure of 4.5 MPa. Grouting operation did not cease until the permeability of the curtain reached $q < 5$ Lu.

4.3. Connection to concrete face slabs

If the toe plinth of a concrete face rockfill dam is built on overburden layers and a concrete wall is used to cut off the underground seepage, then a reliable connection between the face slabs

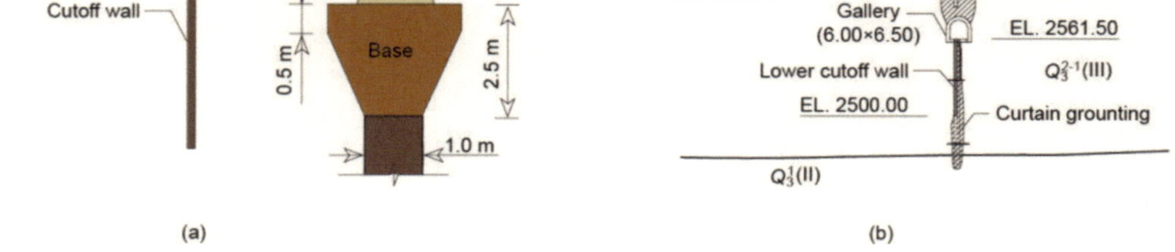

Figure 9. Connection techniques in the Yele ACRD. (a) Section A-A. (b) Section B-B.

and the cutoff wall should be guaranteed. Now, it is a standard way to use one or two linking slabs to connect the cutoff wall with the toe plinth, as exemplified in **Figure 10**. Water stops are installed at the connection points of different components. The width of the toe plinth and the linking slab(s) should be determined based on the allowable hydraulic gradient of the underlying overburden, and on the permitted three-dimensional displacements that are sustainable for the water stop structures. The designing features of connection systems in typical CFRDs are shown in **Table 4**. In these CFRDs, the width of the linking slab(s) usually ranges from 2 to 4 m.

The watertight structure for the perimetric joints used in most CFRDs (e.g., Aertash) consists of three layers. A "W"-shaped copper water stop is used at the bottom, and a wavy watertight stripe is used as the middle sealer. Plastic filling material is enclosed by a "Ω"-shaped rubber

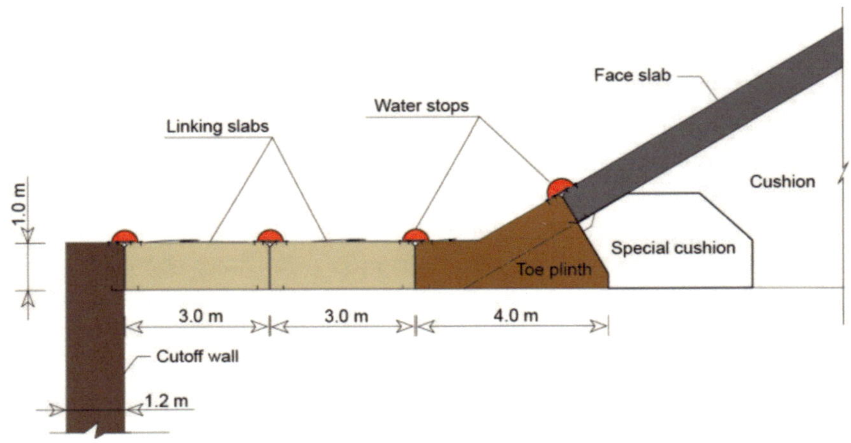

Figure 10. Connection techniques in the Aertash CFRD.

No.	Dam	Dam height (m)	Overburden thickness (m)	Thickness of cutoff wall (m)	Width of toe plinth (m)	Width of linking slab (m)
1	Xieka	108.2	100	1.2	4.0	4.0
2	Nalan	109	24	0.8	8.0	3.0
3	Miaojiaba	111	48	1.2	6.0	3.0
4	Jiudianxia	136.5	56	1.2	6.0	4.0
5	Aertash	164.8	94	1.2	4.0	3.0 + 3.0
6	Chahanwusu	110	47	1.2	4.0	3.0 + 3.0
7	Duonuo	112.5	40	0.8	5.0	3.0
8	Laodukou	96.8	30	0.8	7.0	3.0 + 3.0
9	Jinchuan	112.0	65	1.2	4.0	4.0 + 4.0
10	Gunhabuqile	160.0	50	1.2	4.0	2.0 + 4.0

Table 4. Designing features of the connection systems of typical CFRDs.

plate and fixed at the top of the joints. The gap between the different concrete components is 20 mm, and 12-mm wooden plates are placed in between these components.

5. Foundation improvement techniques

After removing the surficial loose layers, most overburden still requires some additional treatment before using as a dam foundation. Commonly used techniques include compaction, consolidation grouting, vibro-replacement stone columns, high-pressure jet grouting, and so on. These treatment techniques are briefly summarized below.

5.1. Compaction

To provide a firm foundation, vibrating rollers are always used to compact the overburden retained. Dynamic compaction is also commonly used to increase the stiffness and strength of the overburden layers. The authors recommend the Miaojiaba CFRD as an example [9] for dynamic compaction, which was performed before constructing the dam. The average settlement achieved by dynamic compaction was 26.4 cm, and the measured settlement of the overburden during dam operation is about 35–50 cm, indicating that the total settlement may be considerably larger if dynamic compaction has not been performed. Prior to dynamic compaction operations, it is, however, necessary to lower the underground water table.

5.2. Consolidation grouting

Consolidation grouting is always performed to provide a sound foundation for the seepage barrier of dams (e.g., clay core and toe plinth). The depth of grouting generally ranges from 5 to 10 m. The distances between grouting holes and rows range from 2 to 3 m. A concrete plate is usually cast before conducting the consolidation grouting works, serving as a working platform for grouting.

5.3. Vibro-replacement stone column

For weak layers such as sand lenses that are difficult to remove, vibro-replacement stone columns are usually used to densify the soils to form a composite foundation. The stone columns also serve as vertical drainage paths that are beneficial to dissipate the pore pressure within the surrounding soils. In the Huangjinping ACRD [14], vibro-replacement stone columns with a diameter of 1.0 m were set to improve the sand lenses buried more than 25 m below the ground surface.

5.4. High-pressure jet grouting

High-pressure jet grouting is a ground improvement and soil stabilization method, where a stabilizing fluid is injected at a high velocity into the treated soil under a high pressure. The grouted fluid hardens within the soil, forming well-cemented jet grouted columns. High-pressure jet grouting is a very versatile foundation improvement method and has been used not only in building temporary cutoff wall for cofferdams (e.g., **Figure 1**), but also in treating deeply buried sand lenses within the overburden layers (e.g., **Figure 2**).

6. Summary and conclusion

Great advancements in constructing high rockfill dams on thick overburden layers have been achieved in China over the past 20 years. Successful practice can be attributed to progresses in geological and geotechnical investigation techniques, proper designing and connection of the watertight systems, as well as the careful foundation improvement measures. It can be expected that even challenging geological conditions may be encountered in the future, which poses pressing needs in the following aspects:

a. *Reliable assessment of overburden layers and their engineering properties*. Combined use of traditional and newly invented geological and geotechnical investigation methods may considerably improve the reliability of the proposed results. Design engineers should fully understand the geotechnical parameters at hand and the possible limitations involved.

b. *Numerical simulation techniques*. Computational simulations (such as the finite element method) are playing an increasingly important role in designing. Embedding reasonable and simple constitutive models for dam materials and in-situ overburden soils into a fully coupled procedure can yield reliable predictions on the performance of both dams and their seepage barriers in an economical way. Such constitutive models, however, are still scarce.

c. *Effective emergency countermeasures*. It is very difficult to guarantee completely reliable construction quality for seepage control facilities as they are either enclosed inside the dam or buried deep under the dam. In the event that unexpected leakage does occur anywhere in the dam or foundation, effective countermeasures should be in place to eliminate threatens and to prevent amplification of leakage points.

Acknowledgements

This work is supported by the National Key Research and Development Program of China (No. 2017YFC0404806) and the National Natural Science Foundation of China (Nos. 51779152 and U1765203).

Author details

Zhongzhi Fu[1]*, Shengshui Chen[1] and Enyue Ji[2]

*Address all correspondence to: fu_zhongzhi@yahoo.com

1 Geotechnical Engineering Department, Nanjing Hydraulic Research Institute, Nanjing, P. R. China

2 Key Laboratory of Failure Mechanism and Safety Control Techniques of Earth-Rock Dams, Ministry of Water Resource, Nanjing, P. R. China

References

[1] International Commission on Large Dams (ICOLD). Concrete Face Rockfill Dams, Concepts for Design and Construction. Beijing: China Water Power Press; 2010

[2] Xu Q, Chen W, Zhang Z. New views on forming mechanism of deep overburden on river bed in southwest of China. Advances in Earth Science. 2008;**23**:448-456. DOI: 10.3321/j.issn:1001-8166.2008.05.002

[3] Deng MJ. Advances in key technology for concrete face dams with deep overburden layers under cold and seismic conditions. Chinese Journal of Geotechnical Engineering. 2012;**34**:985-996

[4] Zhang H, Chen J, Hu S, Xiao Y, Zeng B. Deformation characteristics and control techniques at the Shiziping earth core rockfill dam. Journal of Geotechnical and Geoenvironmental Engineering. 2016;**142**(2):04015069. DOI: 10.1061/(ASCE)GT.1943-5606.0001385

[5] Liu SH. Safety assessment of hydraulic structures built on overburden layer. Dam and Safety. 2015;**1**:46-63. DOI: 10.3969/j.issn.1671-1092.2015.01.015

[6] Zhou W, Li SL, Ma G. Assessment of the crest cracks of the Pubugou rockfill dam based on parameters back analysis. Geomechanics and Engineering. 2016;**11**(4):571-585. DOI: 10.12989/gae.2016.11.4.571

[7] Xiong K, He Y, Wu X, Dong Y. Stress and deformation behavior of foundation gallery of Changheba hydropower station. Chinese Journal of Geotechnical Engineering. 2011;**33**:1767-1774

[8] Tian J, Wang P, Liu H, Zhang P. Joint types for impervious walls and core mass of Maoergai core wall rockfill dam. Advances in Science and Technology of Water Resources. 2010;**30**:41-45. DOI: 10.3880/j.issn.1006-7647.2010.04.010

[9] Fang GD, Dang LC. Achievements and main technological problems in constructing dams on thick overburden layers. In: Proceedings of National Conference on Technology for Earth-Rockfill Dam; 15-17 October 2012; Chengdu. Beijing: China Electric Power Press; 2012. pp. 35-46

[10] Li ZX, Li MH. Construction of cut-off wall in deep overburden for Luding hydropower station. Water Power. 2012;**38**:50-53. DOI: 10.3969/j.issn.0559-9342.2012.01.016

[11] Cao XX. Study on seismic safety of high rockfill dam with earth core on the thick overburden layer [thesis]. Wuhan: Wuhan University; 2013

[12] Fan S, Zheng X. Design and construction of foundation anti-seepage for dam of Xiabandi water control project. Water Resources and Hydropower Engineering. 2012;**43**:8-11. DOI: 10.13928/j.cnki.wrahe.2012.10.017

[13] Sun M, Chen J, Chen X. Stress and deformation analysis of the rock-fill dam with deep covering layer by static three dimensional finite element method. Yellow River. 2013;**35**:103-106. DOI: 10.3969/j.issn.1000-1379.2013.06.035

[14] Luo Q, Song W, Zhao X. Construction technology of large thickness cutoff wall with depth over 100 m in Huangjinping hydropower station. Water Power. 2016;**42**:47-50. DOI: 10.3969/j.issn.0559-9342.2016.03.013

[15] Wang W, Höeg K, Zhang Y. Design and performance of the Yele asphalt-core rockfill dam. Canadian Geotechnical Journal. 2010;**47**:1365-1381. DOI: 10.1139/T10-028

[16] Li NH. Recent Technology for High Concrete Face Rockfill Dams. Beijing: China Water Power Press; 2007

[17] Shen T, Li GY, Li Y, Li J, Feng YL. Numerical analysis of joint types between toe slab and foundation of CFRD in alluvial deposit layer. Chinese Journal of Rock Mechanics and Engineering. 2005;**24**:2588-2592. DOI: 10.3321/j.issn:1000-6915.2005.14.030

[18] Lv SX. The design and research of the concrete face rock-fill dam of Jiudianxia hydroelectric project on the Taohe river in Gansu province [thesis]. Xi'an: Xi'an University of Technology; 2004

[19] Deng MJ, Wu LY, Wang Y, Fan JY, Li XQ. Design of dam body structure and seepage control for deep overlying strata in dam foundation of Aertash hydro project. Journal of Water Resources and Architectural Engineering. 2014;**12**:149-155. DOI: 10.3969/j.issn.1672-1144.2014.02.031

[20] Feng J, Xin JS. Design of the concrete face rockfill dam for Duonuo hydropower station. Sichuan Water Power. 2012;**31**:100-102. DOI: 10.3969/j.issn.1001-2184.2012.z1.031

[21] Wu XB, Zhuang JG, Zheng XJ, Liang Q. Design characteristics of face rockfill dam in Laodukou hydropower station. Hydropower and New Energy. 2012;**100**:23-25. DOI: 10.3969/j.issn.1671-3354.2012.01.007

[22] Li WK. Research on drilling for deep overburden layer of Aertash river in Xinjiang. Journal of Water Resources and Architectural Engineering. 2010;**8**:34-53. DOI: 10.3969/j.issn.1672-1144.2010.05.010

[23] U.S. Army of Corps of Engineers (USACE). Geophysical Exploration for Engineering and Environmental Investigations [Engineering Manual]. EM 1110-1-1802; 1995

[24] Goto S, Suzuki Y, Nishio S, Ohoka H. Mechanical properties of undisturbed tone-river gravel obtained by in-situ freezing method. Soils and Foundations. 1992;**32**(3):15-25. DOI: 10.3208/sandf1972.32.3_15

[25] Wang QG, Yan YQ. Engineering properties of the gravel layers in dam foundations of hydropower stations. In: Proceedings of National Conference on Technology for Earth-Rockfill Dam; 27-29 September 2014; Dandong. Beijing: China Electric Power Press; 2015. pp. 343-354

[26] Zhao JM. Large-scale in-situ Tests for the Mechanical Properties of Dam Materials Used in Aertash Concrete Face Rockfill Dam. [Scientific Report]. Beijing: China Institute of Water Resources and Hydropower Research; 2017

[27] Wang L, Liu S, Chen Z, Li Z. Shear strength tests on river overburden in dam site of hydropower station. Water Resources and Power. 2014;**32**(1):122-124

[28] Cheng ZL, Pan JJ, Zuo YZ, Hu SG, Gheng YH. New experimental methods for engineering properties of overburden of dam foundation and their applications. Chinese Journal of Geotechnical Engineering. 2016;**38**:18-23. DOI: 10.11779/CJGE2016S2003

Chapter 2

Seismic Crack Investigation in an Earth Dam by Centrifugal Loading Test

Akira Kobayashi and Akira Murakami

Additional information is available at the end of the chapter

http://dx.doi.org/10.5772/intechopen.78788

Abstract

There are many Earth-fill dams in Japan, which are mostly used for irrigation use. Most of these dams in Japan were constructed by experience over 100 years ago. There are so many irrigation dams, which suffered earthquake damage in the past. Due to the damages, the cracks at the crest in the dam-axis direction have been reported in many cases. For the rock-fill dam recently constructed, the crack on the crest in the dam-axis direction has also been found in the case of a large earthquake. The mechanism of such a crack has not been discussed well. In this study, to clarify the mechanism of a crack in the dam-axis direction, a centrifugal loading test was applied to the dam with a 50 G gravity field. As a result, the critical level of strain was observed at the crest of the model, and it was found that the horizontal displacement at the upper part of the dam was excessive. It can be concluded from the study that the seismic cracks in the dam-axis direction occurred due to the excessive tensile stress, which was not considered in the design process.

Keywords: Earth and rock-fill dam, earthquake damage, crack, centrifugal loading test, image analysis

1. Introduction

In Japan, there are approximately two hundred 10,000 small-Earth dams, which have significant reservoirs for irrigation. Such kinds of dams whose heights are smaller than 15 m are called irrigation dams. Many irrigation dams were constructed by experience over 100 years ago. It is reported that about 20,000 irrigation dams were damaged during the long-term use, which need repairing and strengthening [1].

IntechOpen

© 2018 The Author(s). Licensee IntechOpen. This chapter is distributed under the terms of the Creative Commons Attribution License (http://creativecommons.org/licenses/by/3.0), which permits unrestricted use, distribution, and reproduction in any medium, provided the original work is properly cited.

Many irrigation dams were not designed for earthquake resistance, and the cracks at the crest in the direction of the dam axis are remarkable. **Figure 1** shows the example of the crack at the crest of the Earth-fill dam. The relatively big open crack occurred. This type of open crack can be seen in the damaged Earth-fill dam so often. Not only the old Earth-fill dam but also the recent rock-fill dam has a crack in the dam-axis direction when the big earthquake hit. **Figure 2** shows the crack situation at the crest of the rock-fill dam constructed in 1988. This dam was designed for earthquake resistance. Although the acceleration was not measured, it was inferred that relatively big acceleration occurred because the earthquake inducing the crack was in 2011 off the Pacific coast of Tohoku. The open crack was propagated at a 3 m depth from the crest. The cracks shown in the figures cannot be seen to be induced by shear failure. The earthquake resistance is examined for shear failure with the slip-circle method according to the standard. It is difficult to

Figure 1. Crack at the Earth-fill dam.

Figure 2. Crack at the rock-fill dam.

repair the damaged dam effectively when the mechanism of the crack on the crest is not clarified. On the contrary, if the mechanism becomes clear, an effective counterplan may be considered.

In this chapter, the centrifuge loading test is carried out to confirm the mechanism of the crack at the crest. Although the authors already carried out the 1-G shaking test and inferred that the tensile stress is the reason for the crack by observing the behavior of the cross section [3], this study aims to confirm more concretely the reason. Therefore, the centrifuge loading test of 50 G was planned to simulate the more realistic situation and the observation from the vertical direction also tried to observe the crack situation. Moreover, the simple numerical simulation has tried to reproduce the experimental results.

2. Previous studies

The 1-G shaking table tests were carried out to investigate the dam and embankment behavior, and it was found that the acceleration response at the upper part of the embankment has a large effect on the behavior of the slope [3]. The crack in the dam-axis direction was also examined and the crack was considered to be caused by tensile stress. The tensile stress was affected by the vertical vibration as well as horizontal one [4]. Like others, the relation between the natural period and failure feature was investigated [5]. The effect of the aspect ratio of the dam on the vibration mode was also examined [6].

For the centrifugal loading tests, the acceleration response and residual deformation were investigated for the Earth-core rock-fill dam and concrete-faced rock-fill dam [7]. The effect of the liquefaction of foundation on the deformation of the dam was examined [8]. Moreover, the seismic response and liquefaction of loose embankments were also investigated [9].

As mentioned earlier, the previous studies focused on the seismic response, slope sliding and deformation and so the situation of the surface of the dam was focused. On the other hand, the authors investigated the behavior of the cross section by 1-G shaking table test [2] as mentioned earlier. **Figure 3** is the strain distribution of the model used for 1-G shaking table test. By using

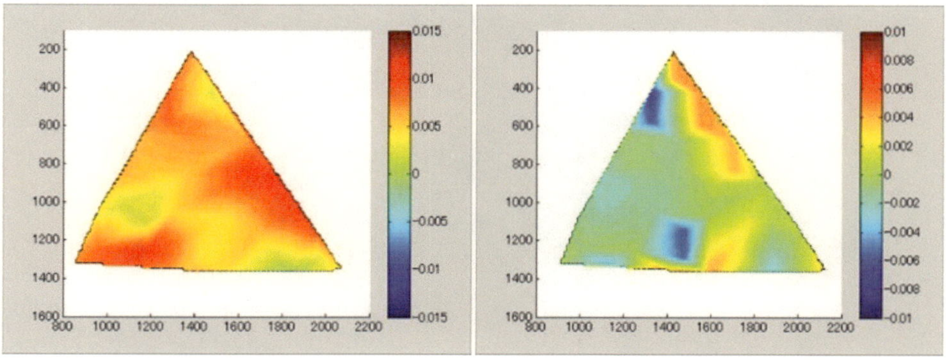

Figure 3. Strain distribution by 1-G shaking test [2]: (a) shear strain and (b) volumetric strain.

image analysis, the behavior of cross section was observed. It was found that the share stain became large at the slope, and the large volumetric strain was observed at the upper part of the dam. In this study, a similar image analysis is used for the centrifugal loading test.

3. Centrifugal loading test

3.1. Test conditions

The centrifugal loading tests were conducted under 50-G with a 1/50 scaled model. In this experiment, the model with a height of 100 mm and upstream and downstream gradients of 1:1 was used. **Figure 4** shows the schematic view of the model. The model was made from

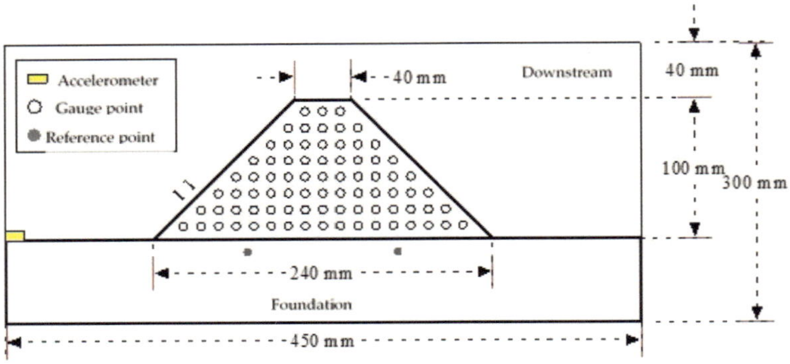

(a)

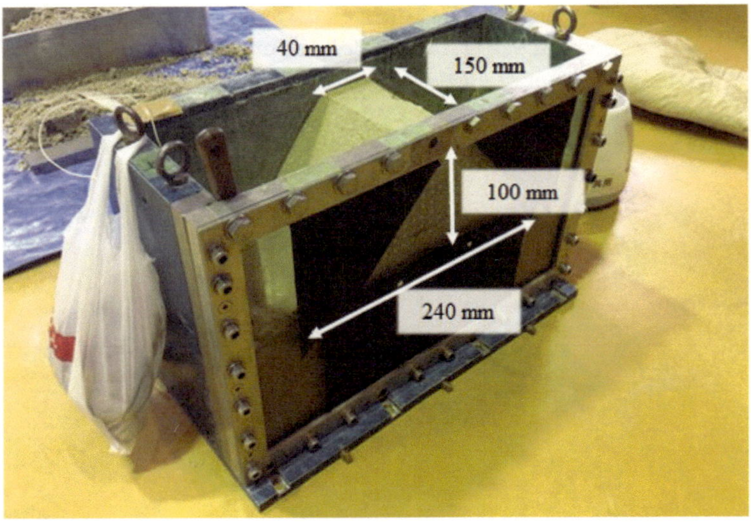

(b)

Figure 4. Schematic view of the model and soil box for centrifugal loading test. (a) Schematic view of the model; and (b) the situation of soil box for centrifugal loading test.

No. 7 silica sand and kaolin clay with the mixture ratio of 8:2 by dry weight. **Figure 5** shows the particle distribution of the mixture. The water content of the mixture was 13%. In order to evaluate the seismic behavior of the model, the gauge points were placed on the surface of the embankment model and the reference points were set at the foundation. The total number of the points is 82. To reduce the friction between embankment model and wall of the soil box, the silicone grease was painted on the wall surface. The accelerometers were installed on the foundation. The foundation was made of rigid material. By assuming that the distance of the reference points on the rigid foundation is not changed, the coordinate of the gauge points is calibrated.

The model was excited with a horizontal sine wave of 50 Hz whose amplitude is 1.5 mm. The input seismic wave corresponds to the earthquake ground motion with a peak acceleration of about 300 gal, and the frequency is 1 Hz.

3.2. Digital image analysis

In order to evaluate the displacement of the gauge points, the digital image analysis method used in the 1-G shaking table test [2] is applied.

Firstly, the static image is taken before the model is tested. While the model is excited, continuous images are taken by the high-speed camera of which the shooting speed is 1000 fps. The images are transformed into black and white binary images. The noise reduction is, then, carried out as shown in **Figure 6**. The number of the points is confirmed to be 82 at this stage. Then, the coordinates of the gauge points are measured in the unit of pixels by calculating the center position of each white element representing the gauge point. Finally, the distance between two reference points at the foundation, which was precisely 150 mm, is measured in the unit of pixels. The scale calibration is carried out by the distance of reference points and the coordinate of the gauge points is calculated as the relative location of the reference point in the unit of mm. By repeating this procedure for all dynamic images, the displacement variation of each gauge point can be obtained.

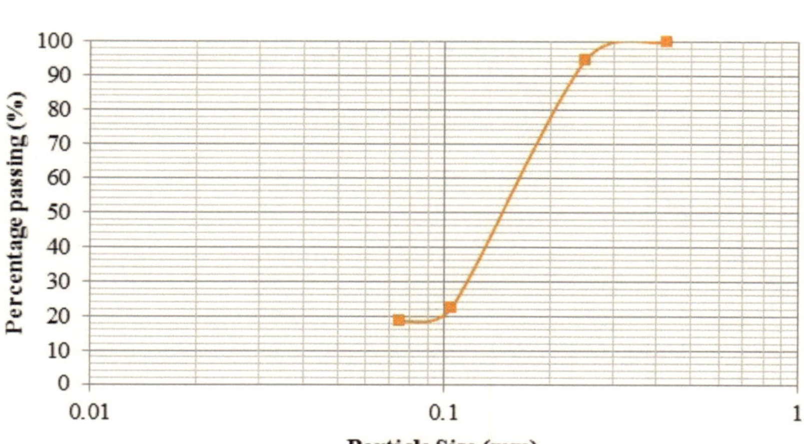

Figure 5. Particle size distribution used for the embankment model.

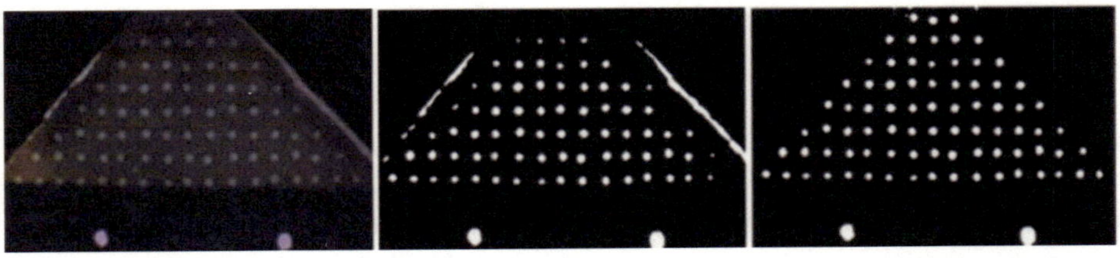

Figure 6. Examples of image analysis process: (a) picture image; (b) binary image; and (c) noise reduction result.

3.3. Calculation of strain

As the gauge points are regarded as the nodes, the cross section is discretized with a triangular element. By using the plane strain condition, the shear and volume strains are calculated for each element with the theory of the finite element method.

4. Experimental results

4.1. The behavior of cross section

By using the method mentioned above, the behavior of the cross section is examined. **Figure 7** shows the final displacement vector distribution after experiment and the situation of cracks. The cracks occurred at the upper part of the dam body of which the situation is explained later. The downward displacement vector is large at the upper part while the displacement at the lower part is minimal.

The gauge points shown by the number in **Figure 8** are focused on highlighting the deformation pattern. The downward displacement at the vertical centerline of the dam body, of which

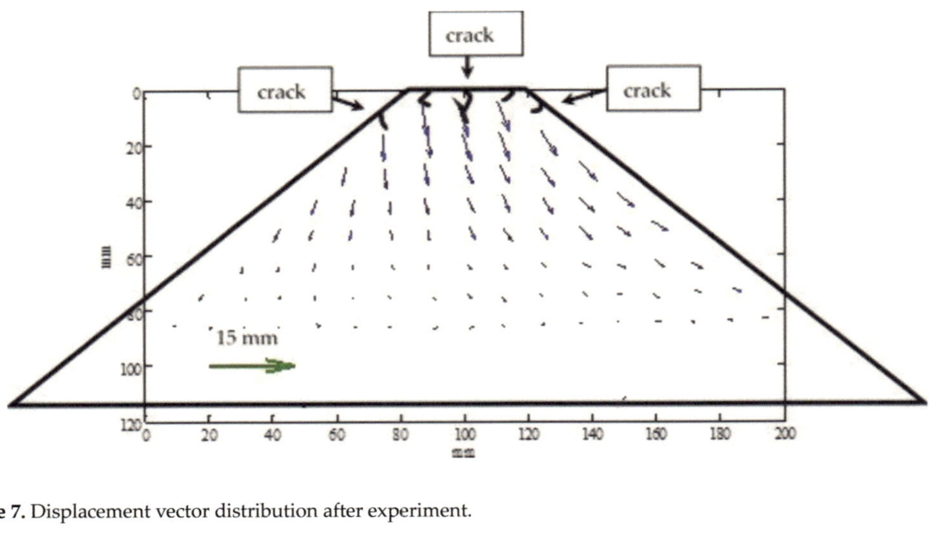

Figure 7. Displacement vector distribution after experiment.

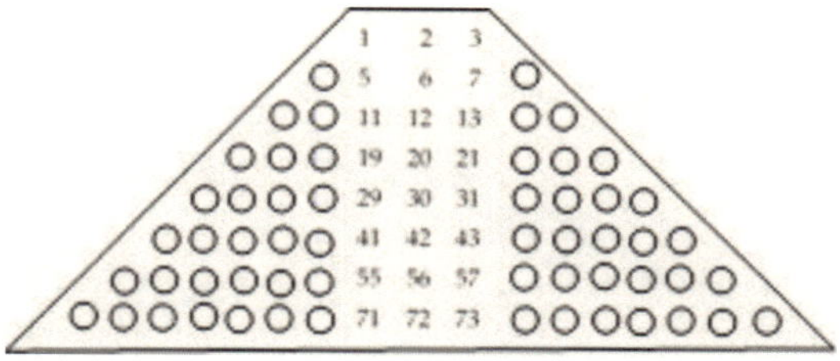

Figure 8. Number of gauge points highlighted for observation.

the number is from 2 to 72, is shown in **Figure 9**. To observe the tendency of the settlement, the displacement is averaged with the period of 0.02 s which is the same as the period of shaking. The downward displacement becomes gradually large with the height of the gauge point. The deformation continues during the experiment at the upper part while the displacement at the lower part does not change so much during the experiment. It is found that the deformation of the upper part occupies the settlement of the dam.

Then, to observe the horizontal deformation, the change in the horizontal distance between two gauge points is investigated. For example, at the upper most part of the dam, the horizontal distance of gauge points 1 and 3 is presented (see **Figure 8**). **Figure 10** shows the change in the horizontal distances at various heights of the dam with time, in which the legend means the number of gauge points used for the distance calculation. While the horizontal distance becomes large at the height of the middle, the number of 29–31, at the early stage, the ones at the upper parts, the number of 1–3 and 5–7, gradually increase with time. The upper parts have a significant change in the horizontal distance finally. Entirely, the center part of the dam has the horizontal tension behavior except for the lowest part.

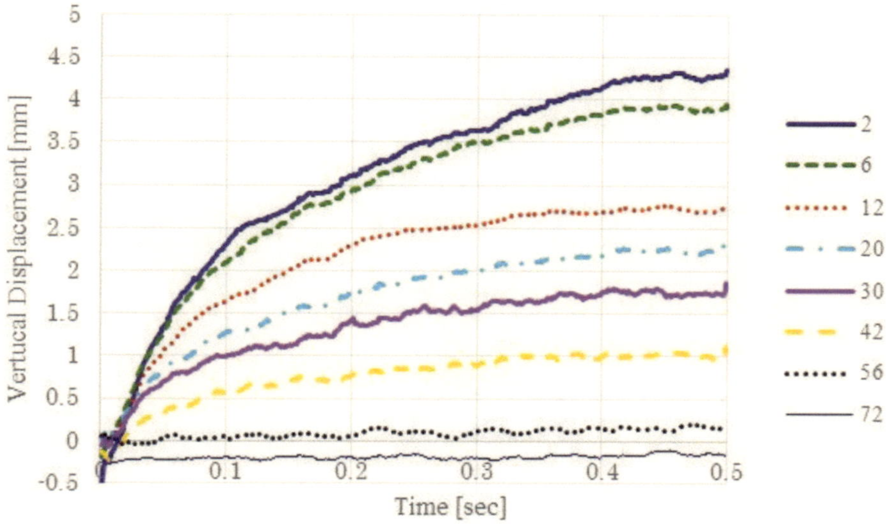

Figure 9. Downward displacement at the centerline of the dam.

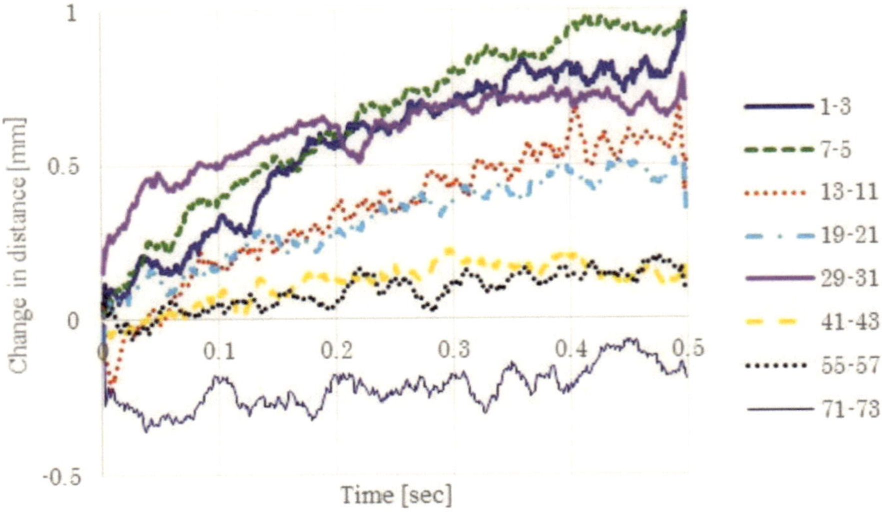

Figure 10. Change in the horizontal distance at the centerline of the dam.

Figure 11 shows the change in shear strain distribution. The situation starts from the left end of shaking, moves to the center and right end, and then reverses to the center and finally returns to the left end. The positive value means the shear strain at downstream shown in **Figure 4(a)**, which coincides with the right-hand side. The time

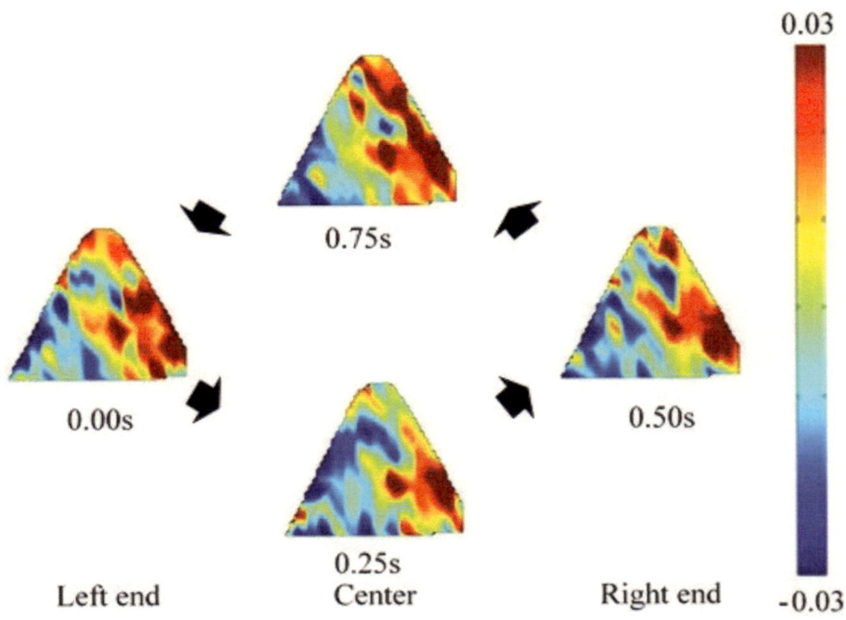

Figure 11. Change in shear strain distribution.

http://dx.doi.org/10.5772/intechopen.78788

means the corresponding actual one. The shear strain to upstream direction occupies the upstream lower part, while the downstream side of the dam shows the shear strain at the downstream direction. The shear strain in both directions becomes larger at the lower part than the upper part.

Figure 12 indicates the volumetric strain distribution change. The positive value means the extension. It is found that the extension occurs at the upper part during the shaking. At the lower part, the extension and compression distribute apparently, and so the striped pattern can be seen. It is shown that the part showing the extension strain coincides with the part indicating the horizontal tension behavior explained (see **Figure 11**).

4.2. Behavior of crest

The situation of the crest after the test is shown in **Figure 13**. The crack is marked with the red line. The crack in the dam-axis direction can be seen at the center of the crest as shown at the actual damage site. The sliding behavior cannot be found at the slope of the dam. In order to investigate the deformation situation at the crest, an additional experiment is carried out, at which the gauge points are set at the crest and slope as shown in **Figure 14**. As the situation was photographed from the top, the strain is estimated just for the horizontal plane (**Figure 15**). Therefore, the obtained strain is not the one on the slope but the apparent strain on the horizontal plane. Only the strain distribution at the crest is the accurate value as a strain.

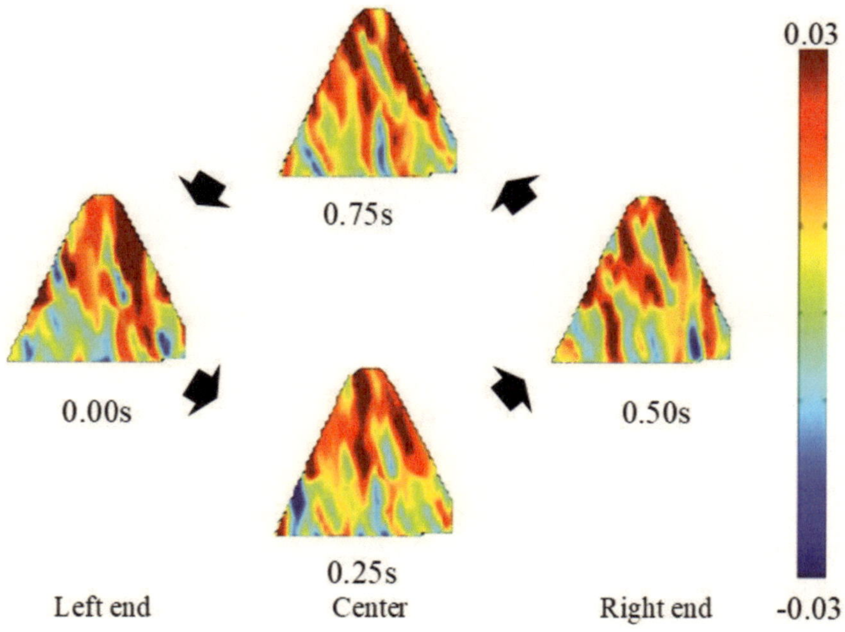

Figure 12. Change in volumetric strain distribution.

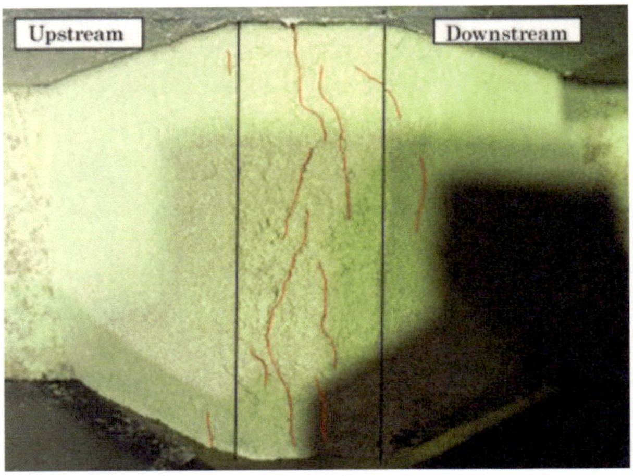

Figure 13. Cracks at the crest after the experiment.

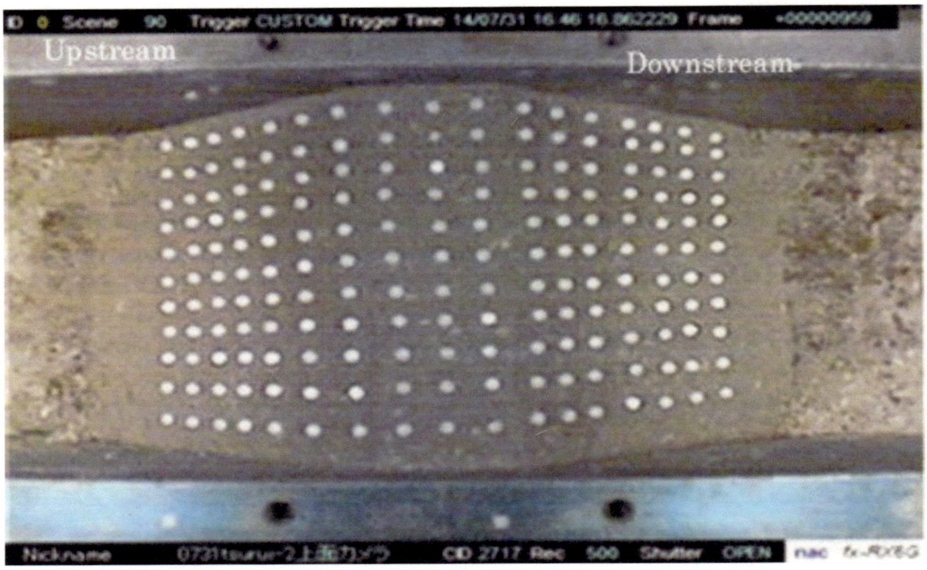

Figure 14. The gauge points on the crest and the slope. (a-1) Left end: shear strain; (a-2) left end: volumetric strain; (b-1) center: shear strain; (b-2) center: volumetric strain; (c-1) right end; shear strain; (c-2) right end: volumetric strain; (d-1) center: shear strain; (d-2) center: volumetric strain.

The shear strain is shown to develop to the upstream and downstream direction near the wall of the sand box. These results may be caused by the friction between wall and dam model. The direction of shear strain distributes alternatively, and the large shear strain develops in the upstream and downstream direction.

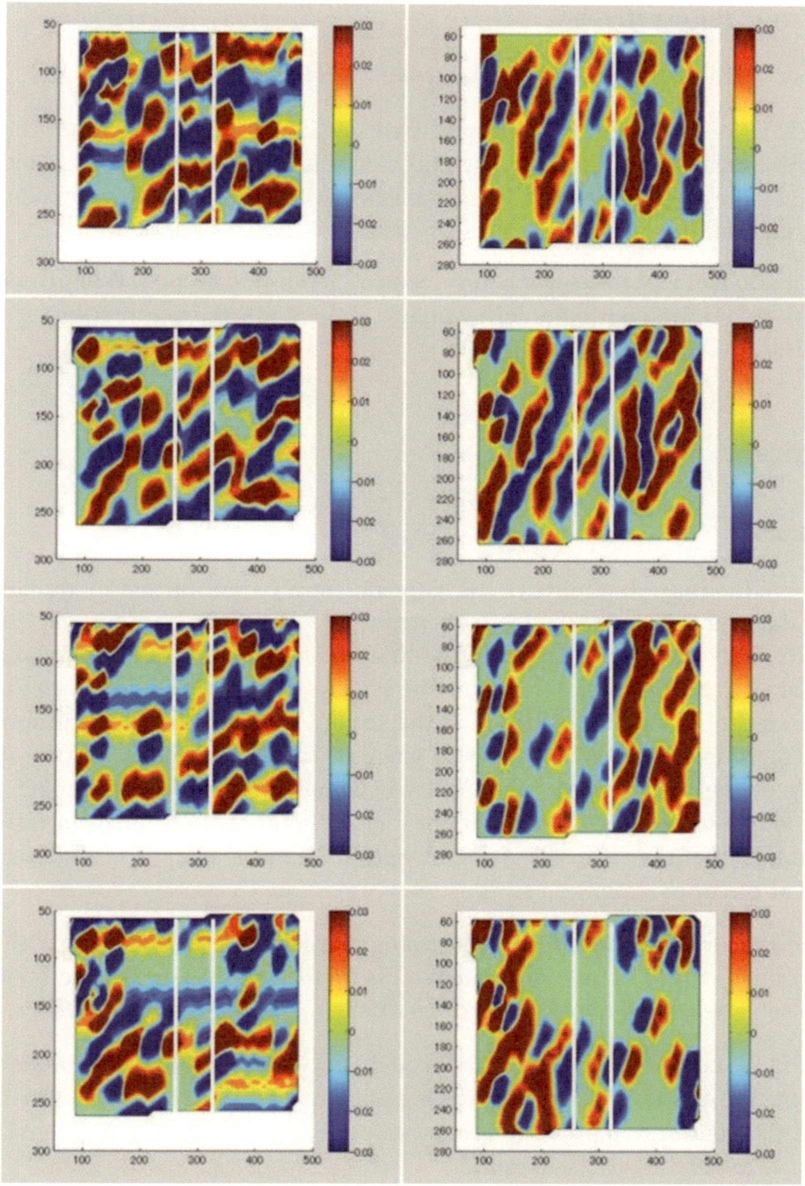

Figure 15. Shear and volumetric strain at the crest and slope.

On the other hand, while the volumetric strain also shows the stripe pattern, the direction of the strain development is dam-axis one, which coincides with the crack situation shown in **Figure 13**. It can be found that the crack shown in **Figure 13** is caused by the extension strain at the crest.

4.3. Summary of the experiment

It can be concluded from the experimental results as follows:

1. The cracks in the dam-axis direction can be realized by the experiment. Such a crack shown at the earthquake damage is caused by the extension strain. Therefore, the extension stress is caused by the parts of the crest. While the extension failure is not examined in the design process, the counterplan may be necessary as the earthquake resistance.
2. The shear strain at the cross section developed to the slope direction. While the shear strain distribution coincided with the sliding failure of the slope, the failure form by sliding was not observed.

5. Numerical examination

5.1. Conditions

To investigate the reproducibility of the extension strain distribution by numerical simulation, the dam model of which size is the same as the 1-G shaking test [2] is simulated. In this study, two cases are examined. The first one is the case in which the dam body is directly shaken at the bottom of the dam. The second one is the case in which the thin base is added under the dam. The objective of the second case is to input the inhomogeneous wave into the dam body. When a given wave is subjected to the bottom of the base, the shaking of the bottom of the dam body becomes a little inhomogeneous, while the amplitude is not so different. The dam body has a height of 150 mm and upstream and downstream gradients of 1:0.545. As the slope gradient is steep, the shaking would be intensified at the upper part.

To examine the strain distributions, the elasto-plastic finite element method (FEM) using Mohr-Coulomb's criteria as the yield function is applied in this study. The bottoms of the models are subjected to 2.4 Hz of a horizontal sine wave. The amplitude is 280 gal.

Figure 16 shows the schematic view of the models and **Table 1** indicates the material properties. By setting the elastic modulus of the base a little smaller than that of the dam, the shaking situation at the bottom of dam body becomes inhomogeneous by the different behavior of reflected wave depending on the location.

5.2. Results

Figure 17 shows the shear strain distribution and **Figure 18** indicates the volumetric strain distribution at the time when the model locates the center of shaking from downstream to upstream. It is found that the large shear strain occurs at the lower part for both cases while the direction is different for the case. Although Case 1 shows that the shear strain to the same direction occurs, the shear strain to the different direction is distributed for Case 2. For the volumetric strain, the maximum value is shown at the top and both ends of the bottom, and

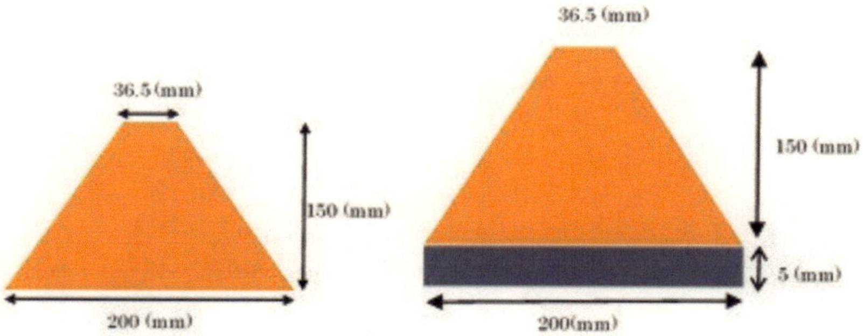

Figure 16. Model size of the numerical simulation. (a) Case 1 and (b) Case 2.

Parameters	Dam body	Base (Case 2)
Cohesion (kPa)	15	15
Unit weight (kN/m³)	18	18
Elastic modulus (kPa)	3000	2850
Internal friction angle (°)	35	35
Poisson's ratio	0.3	0.3

Table 1. Material parameters used for numerical analysis.

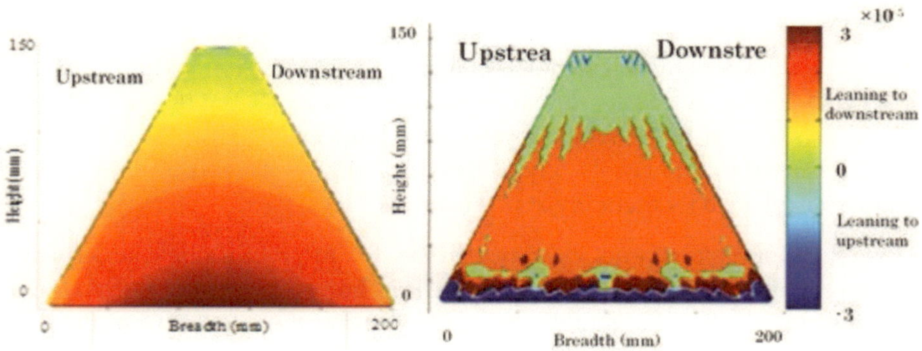

Figure 17. Shear strain distribution of the numerical simulation. (a) Case 1 and (b) Case 2.

the upstream side has the compression strain, and the downstream side has the extension strain for Case 1. On the other hand, Case 2 shows that the extension and compression strain distribute alternatively. This stripe pattern can be seen in the centrifugal loading test as shown in **Figure 12**. However, the large extension strain cannot be seen in the numerical simulation.

As a result of the numerical simulation, the analysis results are different from the experimental results. In particular, the large extension strain at the upper part cannot be realized by the

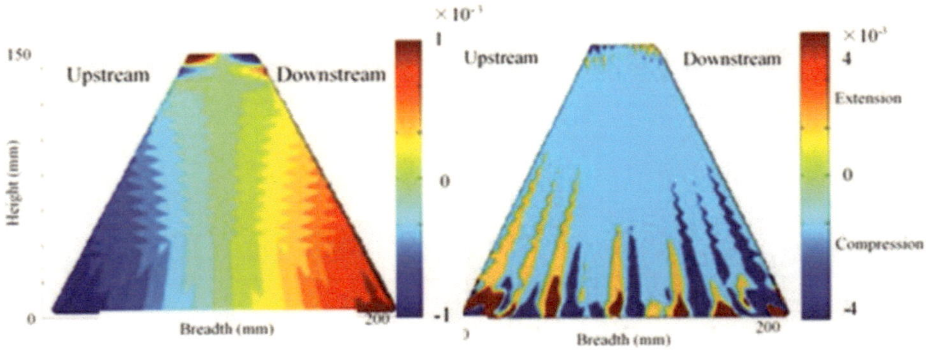

Figure 18. Volumetric strain distribution of the numerical simulation. (a) Case 1 and (b) Case 2.

analyses. The distribution of the shear strain is also quite different from the observed one. As a present conclusion, the ordinary elasto-plastic model is not suitable for the dynamic analysis of the dam.

6. Discussions

Many Earth dams had the cracks in the direction of dam axis at the crest when a big earthquake happened. In this study, to examine the mechanism of the cracks along the dam axis, the centrifugal loading tests are conducted. Moreover, numerical analyses are used to examine the mechanism theoretically.

It was found from experimental examination that the cracks in dam-axis direction shown as the earthquake damage were caused by the extension strain. Therefore, the extension stress was caused by the parts of the crest. However, the large extension strain at the upper part could not be realized by numerical analyses. This means that the cracks in the dam-axis direction cannot be explained theoretically at present.

While the extension failure is not examined in the design process, the counterplan for the cracks is necessary as the earthquake resistance. It is, however, difficult to examine the efficiency because it is impossible that the extension stress at the upper part of the dam body cannot be evaluated properly at present.

The shear strain at the cross section is developed to the slope direction in the experiment. While the shear strain distribution coincided with the sliding failure of the slope, the failure form by sliding was not observed. On the other hand, the distribution of the shear strain by numerical analysis was quite different from the observed one. Although the earthquake-resistance is examined for shear failure by Newmark's method [10] or the method by Watanabe and Baba [11] in Japan, the discussion about the critical situation of the seismic behavior of the fill dam would be needed.

Acknowledgements

This study was carried out with the support of the grant in aid for scientific research (A) and (B) of JSPS. The experiments were assisted by Mr. Tsurui and Mr. Sugano who were the graduate students of Kyoto University at that time. We appreciate the financial and technical support.

Author details

Akira Kobayashi[1]* and Akira Murakami[2]

*Address all correspondence to: koba5963@kansai-u.ac.jp

1 Department of Civil, Environmental and Applied systems Engineering, Kansai University, Osaka, Japan

2 Kyoto University, Kyoto, Japan

References

[1] Kato T. Flood mitigation function and its stochastic evaluation of irrigation ponds. Bulletin of the National Research Institute of Agricultural Engineering. 2005;(44):1-22 (in Japanese)

[2] Miyanaga Y, Kobayashi A, Murakami A. 1-G model test with digital image analysis for seismic behavior of earth dam. Geotechnical Engineering Journal of the SEAGS & AGSSEA. 2015;**44**(2):27-34

[3] Lin ML, Wang KL. Seismic slope behavior in a large-scale shaking table model test. Engineering Geology. 2006;**86**:118-133

[4] Masukawa S, Yasunaka M, Kohgo Y. Dynamic failure and deformations of dam-models on shaking table tests. In: Proceedings of 13th World Conference on Earthquake Engineering; 2004; Vancouver Canada. Paper No. 2359

[5] Tsutumi H, Watanabe H, Ogata N, Shiomi S. Dynamic test for seismic design of fill-type dam. Tsuchi to Kiso. JGS. 1975;**23**(5):11-20 (in Japanese)

[6] Masukawa S, Yasunaka M, Hayashida Y. Shaking table tests by silicone rubber dam model with different ratio of crest length to dam height. Tsuchi to Kiso. JGS. 2008;**56**(10):16-19 (in Japanese)

[7] Kim MK, Lee SH, Choo YW, Kim DS. Seismic behaviors of earth-core and concrete-faced rock-fill dams by dynamic centrifuge test. Soil Dynamics and Earthquake Engineering. 2011;**31**:1579-1593

[8] Sharp MK, Adalier K. Seismic response of earth dam with varying depth of liquefiable foundation layer. Soil Dynamics and Earthquake Engineering. 2006;**26**:1028-1037

[9] Ng CWW, Li XS, Van Laak PA, Hou DYJ. Centrifuge modeling of loose fill embankment subjected to uni-axial and bi-axial earthquakes. Soil Dynamics and Earthquake Engineering. 2004;**24**:305-318

[10] Newmark NM. Effect of earthquakes on dams and embankments. Geotechnique. 1965; **15**(2):139-159

[11] Watanabe H, Baba K, Hirata K. A consideration on the method of evaluation for sliding stability of fill dams during strong motion earthquakes based upon dynamic analysis. CRIEPI Research Report. No. 381020. 1981 (in Japanese)

Chapter 3

Three-Dimensional CFD Simulations of Hydrodynamics for the Lowland Dam Reservoir

Aleksandra Ziemińska-Stolarska

Additional information is available at the end of the chapter

http://dx.doi.org/10.5772/intechopen.80377

Abstract

This chapter deals with the processes by which a single-phase 3-D CFD model of hydrodynamics in a Sulejow dam reservoir was developed, verified, and tested. A simplified volume of fluid (VOF) model of flow was elaborated to determine the effect of wind on hydrodynamics in the lake. A hexahedral mesh with over 17 million elements and a k-ω SST turbulence model were defined for single-phase simulations in steady-state conditions. The model was verified on the basis of the extensive hydrodynamic measurements. Excellent agreement (average error of less than 10%) between computed and measured velocity profiles was found. The simulation results proved a strong effect of wind on hydrodynamics, especially on the development of the water circulation pattern in the lacustrine zone in the lake.

Keywords: CFD model, hydrodynamics, dam reservoir, Sulejow reservoir

1. Introduction

In recent years, numerical models of flow in water bodies are widely used to provide an accurate description of flow velocity distribution. Although all the hydrodynamic processes going on in an aquatic system are difficult to describe, some authors have reviewed the main mechanisms, from large scale [1–5] to small scale and mixing processes [6, 7] or both [8, 9]. Three-dimensional numerical models of flow calculations in natural water bodies have been successfully applied for rivers with single complex flow features [10–14] although there is a little experience with modeling of flow in large dam reservoirs [15]. Studies of flow through curved or straight, open channels of simple cross-sectional shape (rectangular or trapezoidal) have been reported by Demuren [16] and Meselhe [17]. Both authors presented a 3-D

IntechOpen

© 2019 The Author(s). Licensee IntechOpen. This chapter is distributed under the terms of the Creative Commons Attribution License (http://creativecommons.org/licenses/by/3.0), which permits unrestricted use, distribution, and reproduction in any medium, provided the original work is properly cited.

numerical model, for the calculation of turbulent flow in meandering channels. The works provided a hydrodynamic basis for the study of the mechanisms for the formation of river meanders. The authors employed a finite volume numerical method to solve the full Reynolds-averaged Navier-Stokes (RANS) equations in conjunction with the standard k-ε model. The case exhibits some of the features encountered in real rivers, including longitudinal curvature and varying bed topography. Demuren's results were in reasonable overall agreement with the mean velocity measurements of Almquist and Holley [18]. Meselhe focused on proposing a simplified approach for calculating the water surface elevation as a part of the overall solution procedure.

All previous studies with both straight and curved channels had adopted a rigid-lid assumption for modeling the free surface. The rigid-lid approximation is a commonly used simplification in the study of density-stratified fluids mainly in oceanography. Assumes, that the displacements of the surface are negligible compared with interface displacements. In a research conducted by Sinha [11], a 3-D model of flow through a 4-km stretch of the Columbia river, downstream of the Wanapum Dam, was developed and validated. Authors succeeded in modeling of flow in rapidly varying bed topography and the presence of islands. In the study three-pronged strategy—composed of the field measurements, the laboratory experiments, and the numerical model—was undertaken. The model solves the (RANS) equations closed with the standard k-ε turbulence model.

In today's development level of hardware and software, increasingly complex, engineering tasks can be managed, and, thus, three-dimensional hydrodynamic modeling of dam reservoirs became a potential tool for determining more accurate picture of the dependencies, which prevail in the aquatic ecosystem. Whereas in most studies one- and two-dimensional description of flows may be sufficient, in some special cases, three-dimensional approach is needed to determining, for instance, flow patterns in bends or in the vicinity of hydraulic structures (dams and weirs). These issues are evidently three-dimensional, and spatial character directly affects pollutant transport processes.

Hydrodynamic processes determine the movement of suspended and dissolved matter; heat transfer; intensity of circulation inside the ecosystem; the speed of contaminating processes; and the self-purification of the reservoirs, ultimately provide conditions of the ecosystem function.

Kennedy et al. [15] applied a 3-D model to estimate the hydraulic residence time (HRT) for the Thomas Basin (part of the Wachusett Reservoir in central Massachusetts). The basin was modeled using the FLUENT software package with particles used to track travel time in steady-state conditions. A tetrahedral mesh was used with accurate description of basin bathymetry. The model solved the transient Reynolds equations for turbulent flow with standard k-ε closure. Modeling was performed to simulate flow pattern during a period when conditions were isothermal and windless. HRT was estimated to 3–4 days which is about half of the HRT that would be expected based on the theoretical mean residence time. The results of the calculations show that the presence of a primary flow path, large-scale eddies, and stagnation zones contributed to the faster travel time. Reductions in inflow rates produce increased residence times and significant changes in flow patterns. However, the authors did not provide any information about model verification.

In a research conducted by Khosronejad [19], a 3-D CFD model was applied to predict the flow hydrodynamics around power intakes within the Dez dam reservoir (Iran). The study was also devoted to a qualitative analysis of sediment transport at the area around the intakes. For incorporating the effects of turbulent flow, the k-ω model was implemented. The finite volume method was used to discretize the RANS equations. The 3-D single-phase model ran for a limited 320 m long reservoir section. The results show that the flow velocity has a maximum value of 1–2 m/s near power intakes and decreases with distance. In addition, the turbulent intensity increases in the area near intake entrance resulting in increasing bed shear stress near intakes. Khosronejad did not present a test of either coarser or finer mesh resolution; moreover, the model was not verified in the field measurements.

2. Study area

Sulejow reservoir is situated on the middle reach of the Pilica river, which is left-hand side tributary of Vistula river, central Poland (**Figure 1**). One of the biggest artificial reservoirs in

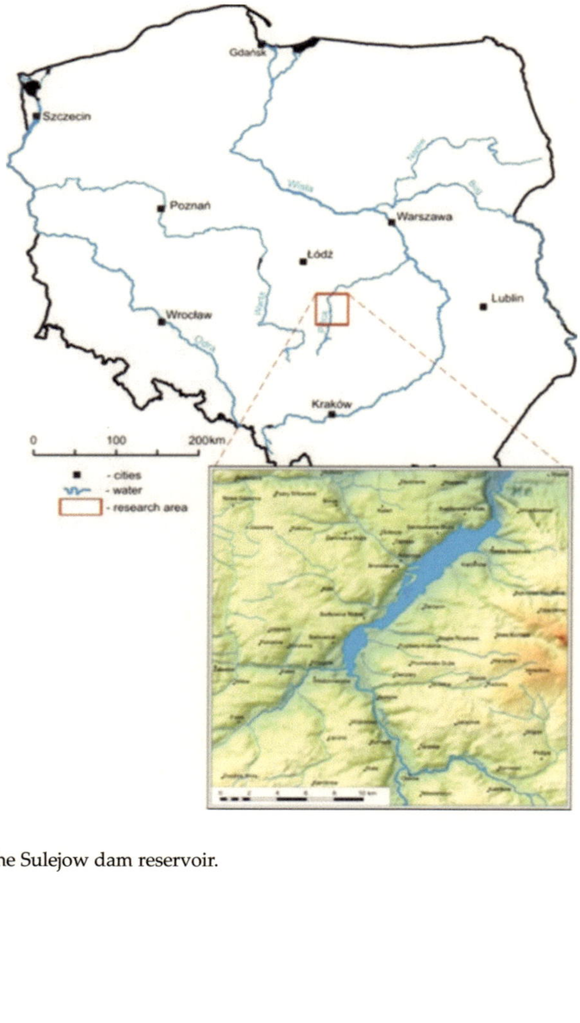

Figure 1. Localization of the Sulejow dam reservoir.

Poland was built by impounding the Pilica river on 138.9 km with a dam in the years 1969–1973. The reservoir is a shallow water body (mean depth 3.3 m) covering a large area (22 km^2).

Sulejow reservoir is a ribbon-type reservoir, which can be divided into two morphological zones each influenced by different forcing agents. The first one (consisting of a riverine zone and a transition zone) is the narrow, shallower part of the reservoir, dominated by the river inflows. The second zone, the wide lacustrine part of the reservoir, is located near the dam. This zone is quite open and behaves like a lake, and the main driving force mechanism causes the movement of water masses wind. The main axis runs from southwest to northeast which is close to the direction of winds that ripple and mix the water. A result is the formation of places with stagnant water on the southern bank of the middle and lower part of the reservoir.

3. Generation of three-dimensional geometry

The first step in the numerical procedure was to prepare the 3-D geometry of the 17-km-long Sulejow reservoir. Gambit Program 2.2.30 (ANSYS, USA) was employed to generate geometry of the water body.

On the basis of the most accurate, available data of the reservoir, geometry was generated using segmentation technique with 36 cross-sectional profiles of the artificial lake. Methodology adopted to determine parameters of the sections was based on the field measurements by using an integrated system: digital depth sounder RESON PC-100 and GPS Trimble 5700. The cross-sectional spacing ranged from 400 to 600 m. **Figure 2** shows an example and locations of the measured sections in the water body.

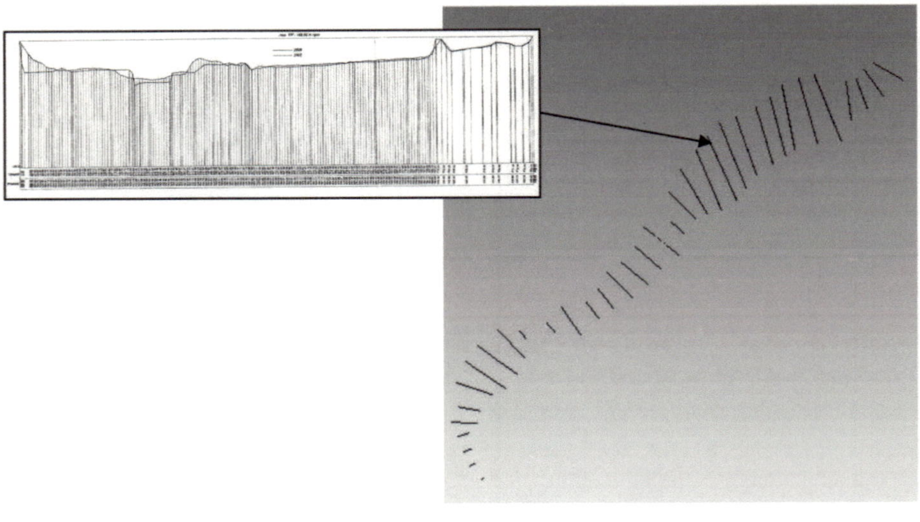

Figure 2. The cross-sectional profiles of the Sulejow reservoir.

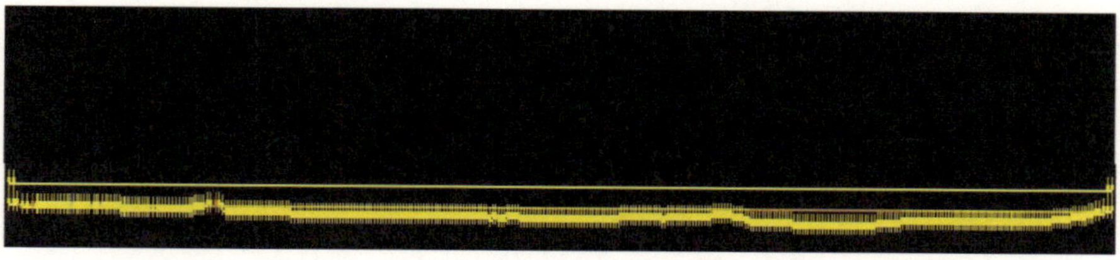

Figure 3. Example of the cross sections generated with Gambit program.

In this study, the "bottom-up" approach was implemented in order to obtain a three-dimensional geometry from the set of separated surfaces. Each measured cross section was defined by the information of the distance between two banks (connection length between probing points in the section was 5 m) and elevation (normal impoundment level in the Sulejow reservoir is 166.6 m amsl). **Figure 3** illustrates an example of the cross section built in the Gambit Program.

The vertices which determine the reservoir bed were connected to form the edges. Subsequently, edges were combined into faces. The last stage consisted in stitching the faces in order to obtain the volumes. After completion of the segmentation procedure, rendering process was conducted, which facilitated generation of three-dimensional geometry of the reservoir, using the faces obtained from the segmentation. **Figure 4** presents the complete geometry that consists of 36 volumes totally.

Terminal berm of the reservoir consists of earth dam and weir, with integrated hydroelectric power station (**Figure 5**). The length of the dam with a weir is 1200 m, the maximum height is 16 m, and the total volume is 567,000 m^3. The jazz is concrete, in the overflow riffle span, middle and left of the weir, drain pipes have been built. Weirs are closed with the oval valves. Cross section of earth dam and weir is shown in **Figure 6a** and **b**. Geometry and computational mesh of the dam are shown in **Figure 7**.

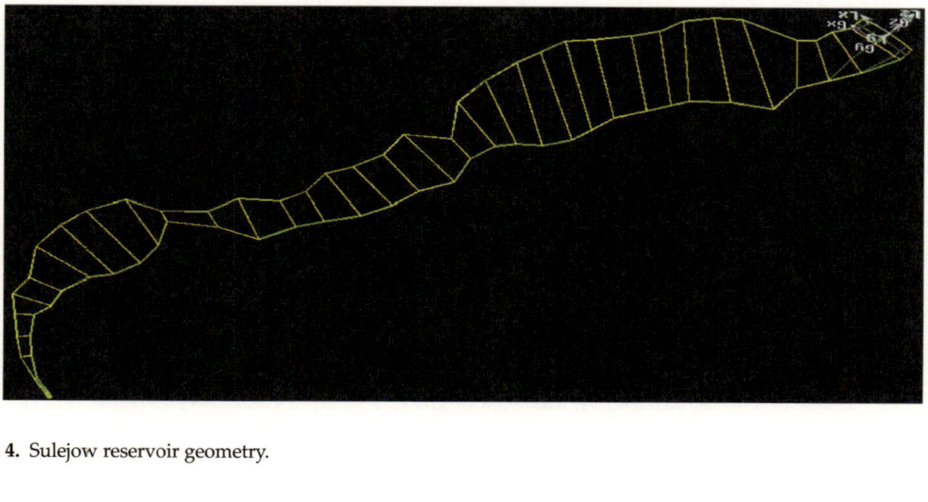

Figure 4. Sulejow reservoir geometry.

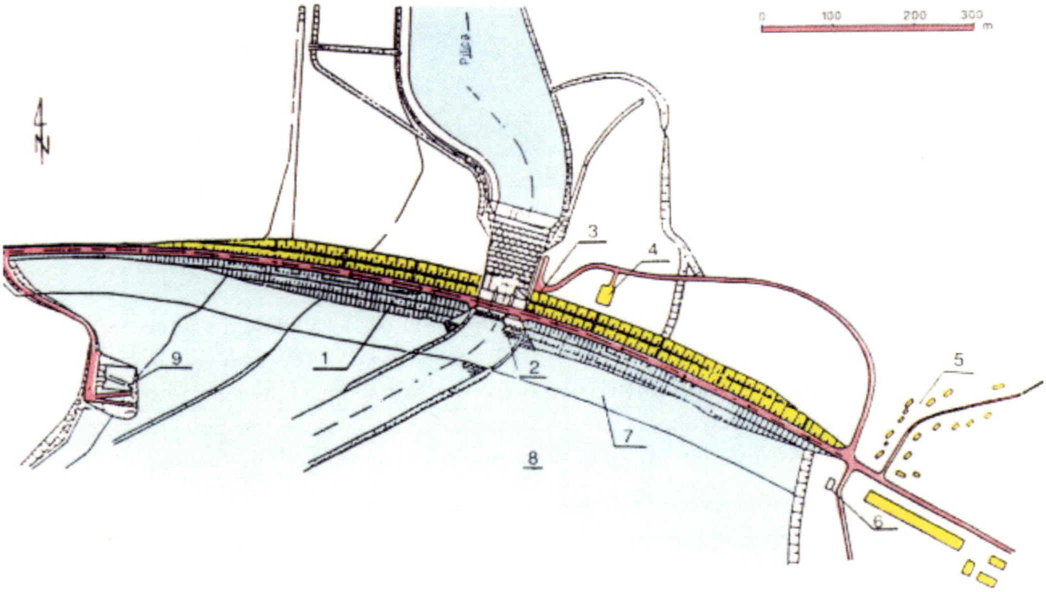

Figure 5. Front step of the weir. (1) Earth dam, (2) weir, (3) electric power station, (4) switching station, (5) operating settlement, (6) buildings, (7) clay band, (8) foil-sealing of the reservoir bottom, and (9) marina.

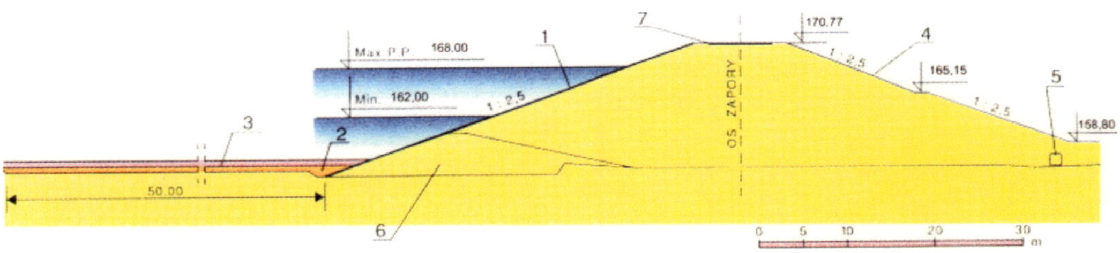

(a)

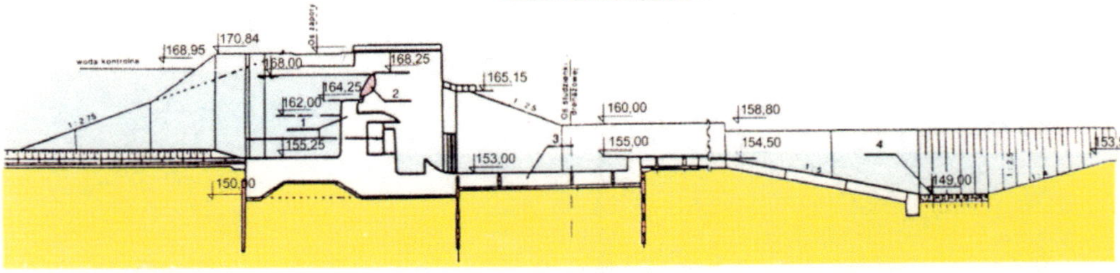

(b)

Figure 6. (a) Cross section of the earth dam. (1) reinforced concrete shield, (2) clay band, (4) grass slope, (5) drainage, (6) embankment, and (7) national road on the dam crest. (b) Cross section of the weir. (1) Elevated sill, (2) oval valve, (3) weir pool, and (4) artificial pothole.

http://dx.doi.org/10.5772/intechopen.80377

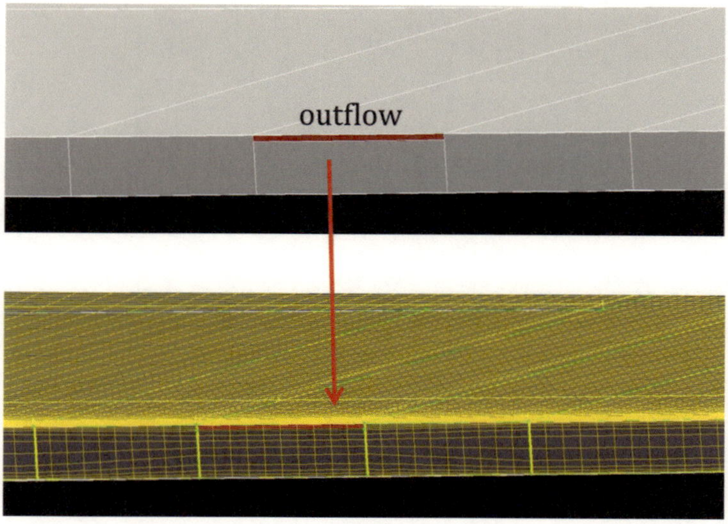

Figure 7. Outflow of the Sulejow reservoir.

The final verification of the accuracy of the reservoir shape was feasible on the basis of satellite photographs, which confirm the correct approximation of the geometry. **Figure 8** shows the satellite image of the Sulejow reservoir with marked cross sections, according to which the 3-D geometry was prepared. Due to the fact that sections were determined every ~500 m, it could not be possible to precisely map the shape of the shoreline; however, this approximation does not affect the nature of the flow throughout the 17-km-long dam reservoir. The generated 3-D geometry of the artificial lake was discretized to perform the numerical solutions.

3.1. Computational mesh

Three-dimensional grid generation for a complex geometry is not a straightforward task, as the grid type has a significant influence upon the quality of the simulations [20]. The crucial issues controlling grid quality are the type of grid, that is, structured and unstructured, grid spacing, and skewness. To apply a finite volume method in CFD model of the reservoir, the 3-D computational domain must be subdivided into a large number of cells. However, the high ratio of the breadth to depth dimensions and irregular shape of the reservoir make this process difficult. Appropriate mesh resolution is linked to the hydrodynamic conditions, the flow features, and the discretization schemes.

In finite volume methods for numerical simulations, the hexahedral mesh (hex mesh) is preferred, as compared to the tetrahedral mesh (tet mesh), owing to the reduced error and smaller number of elements [21]. Typically, tet mesh is preferred for filling irregular spaces, since existing algorithms can semiautomatically subdivide most of the spaces [15]. Structured meshes are better suited to shallow reservoirs, while an unstructured mesh matches better to deeper (or with smaller aspect ratio) reservoirs. On the other hand, generating a hex mesh, with desirable qualities, often requires significant geometric decomposition, which makes the

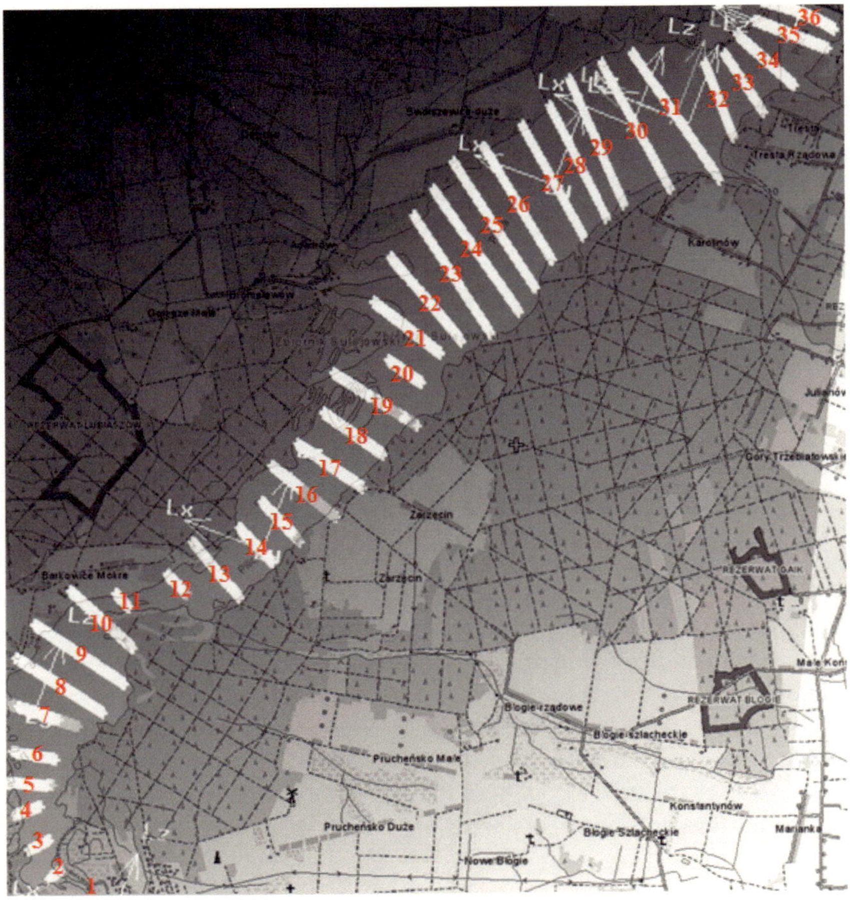

Figure 8. Satellite photograph of the Sulejow reservoir with the cross sections. *Source: geoportal.gov.pl.*

meshing process extremely difficult to perform and automate. As a result, it requires considerable user efforts and may take days or even weeks to develop the proper grid in the case of complex shapes.

To fill the shallow space representing the reservoir area, with minimally skewed, hexahedral cells, the typical cell dimension must be small compared to the depth of the water. Thus, the meshing process becomes computationally expensive due to the requirement of a large number of elements.

While developing the CFD model of flow hydrodynamics, preliminary simulations allowed us to select proper density of the numerical grid, at which a convergent and stable solution can be obtained. In order to generate a mesh for the Sulejow reservoir geometry, the capacity of Gambit 2.2.30 commercial software was used. The domain surface was discretized using structural (hexahedral) mesh with 16,787,820 active cells and 17,717,364 nodes, respectively.

http://dx.doi.org/10.5772/intechopen.80377

The total basin volume was 74,721,600 m^3, which reflects the real value. Four layers of elements across the wall thickness were added to confirm the numerical stability by specifying the height of the first layer, nearest to the wall (0.01), and the growth factor (0.1).

The process of generating structured grid was much more labor intensive than creating an unstructured mesh; however, elaborated model was more stable and converged quickly.

Exclusion from the model of some near-shore regions of the basin, with the shallow depth (<0.3 m) and inconsiderable slope, was necessary to avoid generation of cells with highly acute angles. Removal of strongly skewed elements from the computational domain would have minimal impact on the overall circulation patterns in the basin due to the high ratio of bottom area to the water volume, resulting with minimal flow and very low velocity in this region. This operation removed approximately 10% of the basin surface area but only a small fraction of the basin volume ~1%.

A post-processing check on mesh quality, based on assessing the skewness of the generated cells, indicated that the mesh is of high quality and would not compromise solution stability. The use of a coarser grid (<10^5 and 10^6 elements) caused rapid solution divergence. However, increasing the mesh number to 18×10^6 had no further influence on the results especially on the location of swirl flows. **Figure 9** shows the hexahedral, structured mesh which has been generated for the Sulejow reservoir geometry.

Figure 9. Example of the computational domain with the structural mesh generated in GAMBIT 2.2.30 program. Under magnification, hexahedral elements included in the boundary layer are visible.

Parameter	Values for numerical grids in this work	The limit values in the numerical simulations
y^+	~1.8	≤2
Skewness	~0.8	0–1
Aspect ratio	~1.7	≥1

Table 1. Values of parameter: y^+, skewness, and aspect ratio for the analyzed numerical grid.

The quality of the numerical grid was determined by the shape and the size of the computing field and the total number of elements used in the generated numerical grid and through the position of the first node relative to the plane of the wall. To assess the quality of the numerical grid elements, three parameters were used (Gambit User Guide):

- Parameter y^+—determining the quality of the mesh in the boundary layer
- "Skewness"—determining the quality of the individual grid elements
- "Aspect ratio"—defining the degree of deformation of the mesh elements

Ranges of parameters: y^+, skewness, and aspect ratio for the numerical grid generated in this work are given in **Table 1**.

4. Mathematical model

In order to fully present the hydrodynamics in the dam reservoir, volume of fluid (VOF) model should be developed. Owing to large size and complexity of the computational grid, as well as limited computer capacity, such an approach would be very difficult to apply for a big water bodies. In the literature, VOF technique was adopted only for the CFD modeling of reservoir in a downscale model (1:50) [22].

Sulejow reservoir due to the location, shape of the bowl, and uncovered, flat shores is particularly exposed to the effect of wind. Direction and energy of wind determine the waving movement and mixing of the water, so the impact of the factor, on flow distribution in the CFD simulations, should be taken into consideration.

In the model under wind conditions, the speed, at which the plate was moving on the water surface (which reflects the speed and direction of the wind in the area of the Sulejow reservoir), was determined. For this purpose, independent, 2-D, two-phase problem was resolved. Data concerning wind parameters were provided by the Institute of Meteorology and Water Management in Warsaw and come from the meteorological station (Sulejow-Kopalnia) located near the reservoir. Average speed and direction of wind were selected as 3 m/s and southeast, respectively, based on the daily data from year 2007.

4.1. Simplified VOF CFD model

The 2-D model of a straight channel with a length of 80 m and a width of 6 m (3 m layer of water, 3 m layer of air) was elaborated. **Figure 10** shows the hexahedral, structured mesh which has been generated for the simplified geometry. The grid contains 159,242 elements

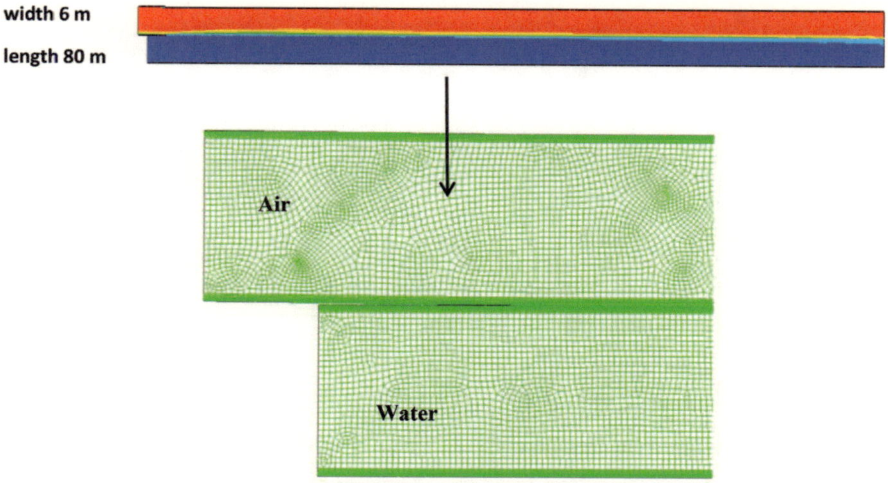

Figure 10. Fragment of the structured, hexahedral mesh generated with Gambit program.

(159,116 hexahedra, 126 wedges) and 321,872 nodes, respectively. A boundary layer was generated consisting of 10 rows.

Wind accelerates surface fluid particles by imparting momentum to the fluid, through surface stresses. In the analysis, the water flows through the air momentum. In the computational domain, the water phase has two outlets that allow for fluid reversing. Analysis of flow in 2-D model was intended to determine which boundary condition best describes the situation that prevails over the water surface by the effect of wind.

4.2. Boundary conditions

For two-dimensional CFD model, the following boundary conditions were imposed:.

Inlet boundary conditions applied for the analyzed domain were two tributaries (Pilica and Luciaza rivers) and outlet (dam) as given in **Figure 11**.

Simulated inflow boundaries were specified with mass flow rates, normal to the boundary. Velocities at each inlet were calculated from the inlet area measurements of stream flow for the Pilica and Luciaza rivers, made by the Regional Board and Water Management in Warsaw in 2007. The monthly values of mass flow rates in the Pilica and Luciaza rivers are presented in **Table 2**.

The definition of the inlet requires the values of the velocity vectors and turbulence properties. For the air inlet, the simulations were first conducted at a speed of 3 m/s. Velocity profile obtained at the outlet of the computational domain was loaded as an input file to receive the velocity profile at the inlet. This approach allows to obtain a fully developed velocity profile for a small domain.

Implementation of volume of fluid model requires an additional boundary condition to be specified, namely, the turbulence intensity at the inlet and turbulent viscosity ratio. Introduction of disturbance into the flow reflects the real features of the flow pattern.

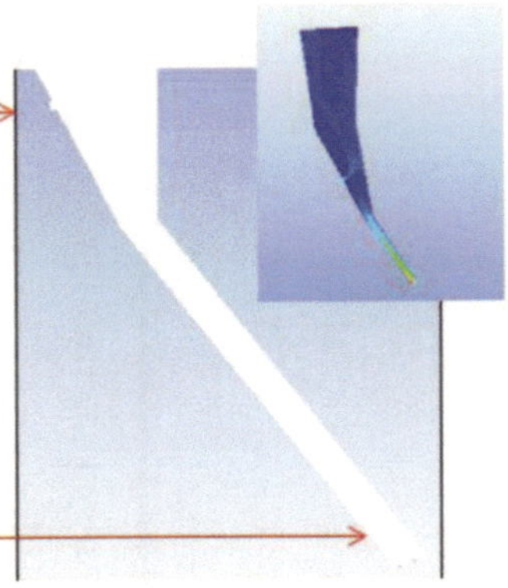

Figure 11. Inlet region in the analyzed domain.

Month	Pilica river (mass flow rate m^3/s)	Luciaza river (mass flow rate m^3/s)
January	19.86	2.80
February	46.31	5.50
March	22.33	4.92
April	23.04	2.19
May	17.36	1.39
June	15.65	1.64
July	16.45	1.91
August	12.83	1.39
September	13.21	1.59
October	16.37	1.69
November	25.50	2.21
December	25.65	2.55

Table 2. The monthly values of mass flow rate in the Pilica and Luciaza rivers in 2007.

The turbulence intensity, I [%], is defined as the ratio of the root-mean-square of the velocity fluctuations $u`$ to the mean flow velocity U_{avg}. The turbulence intensity at the core of a fully developed duct flow can be estimated from Eq. (1):

$$I = \frac{u}{U_{avg}} = 0,16\Re^{\frac{-1}{8}} \tag{1}$$

Kennedy [15] carried out sensitivity analysis on the turbulence intensity in the simple, channel geometry and found that value 4% was able to match the conditions well. The author concluded that this parameter had a noticeable effect on the solution. Following the above findings, turbulence intensity level of 4% was specified at the inlet.

Another parameter, the turbulent viscosity ratio (the ratio of turbulent to laminar (molecular) viscosity), was used as given in Eq. (2). The default value of 10% was applied in the simulations:

$$\beta = \frac{v_t}{v} \tag{2}$$

The air outlet of pressure type was defined in the model. Pressure outlet boundary conditions require specification of a static pressure at the outflow. Convergence difficulties were minimized by specified values for the backflow quantities (backflow turbulence intensity and viscosity ratio). A no-slip boundary condition was applied on the wall.

4.3. Solution methods

Table 3 summarizes the solution conditions and methods used in the modeling process.

4.4. Results of VOF model

Based on the analysis of two-dimensional two-phase model, an appropriate input data for the wall boundary condition, corresponding to real wind conditions at the surface, could be determined. **Figure 12** depicts the velocity profile of water at the outlet of the computational domain. As a result, the replacement velocity of wind, which corresponds to the real, average value in the area of the Sulejow reservoir, was obtained. The mean value of 0.147 m/s was used in the simulations of flow hydrodynamics in the modeled reservoir.

4.5. Results of CFD calculations at wind conditions

Due to the fact that the Sulejow reservoir is a shallow, polymictic lake, the wind will be important for the distribution and mixing of the water masses of two tributaries. The results shown in **Figure 13** confirm this assumption.

Model volume of fluid	
Space	Two-dimensional
Time	Steady
Discretization method	
Gradient	Least squares cell based
Pressure	Body force weighted
Pressure–velocity coupling scheme	Coupled
Momentum	Second-order upwind
Volume fraction	Compressive
Turbulence energy kinetic	Second-order upwind
Specific dissipation rate	Second-order upwind

Table 3. Solution conditions and methods for volume of fluid model.

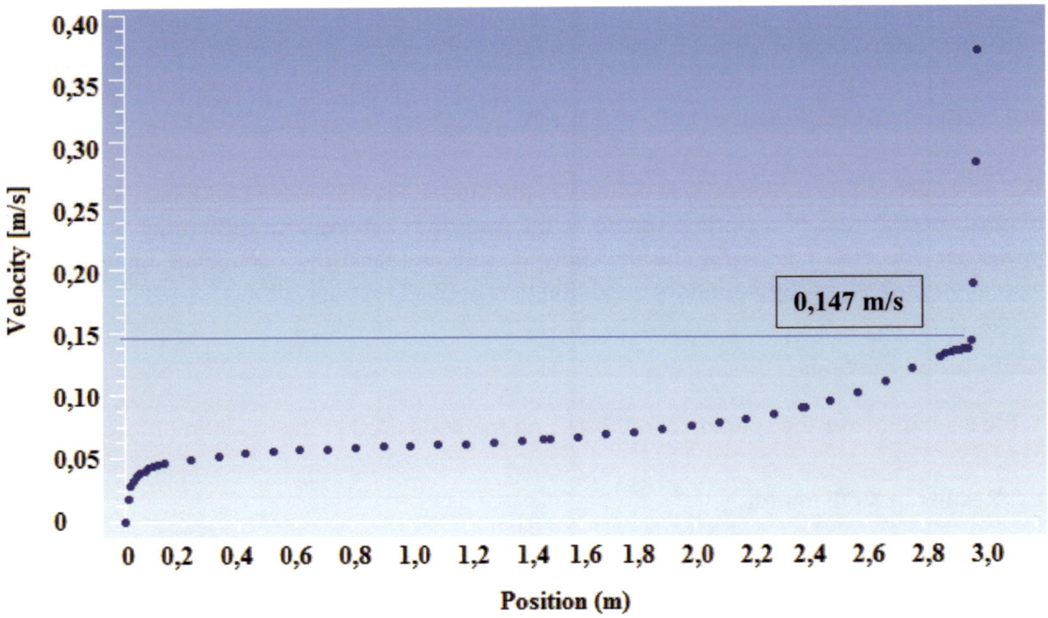

Figure 12. The velocity profile at the interface of water-air.

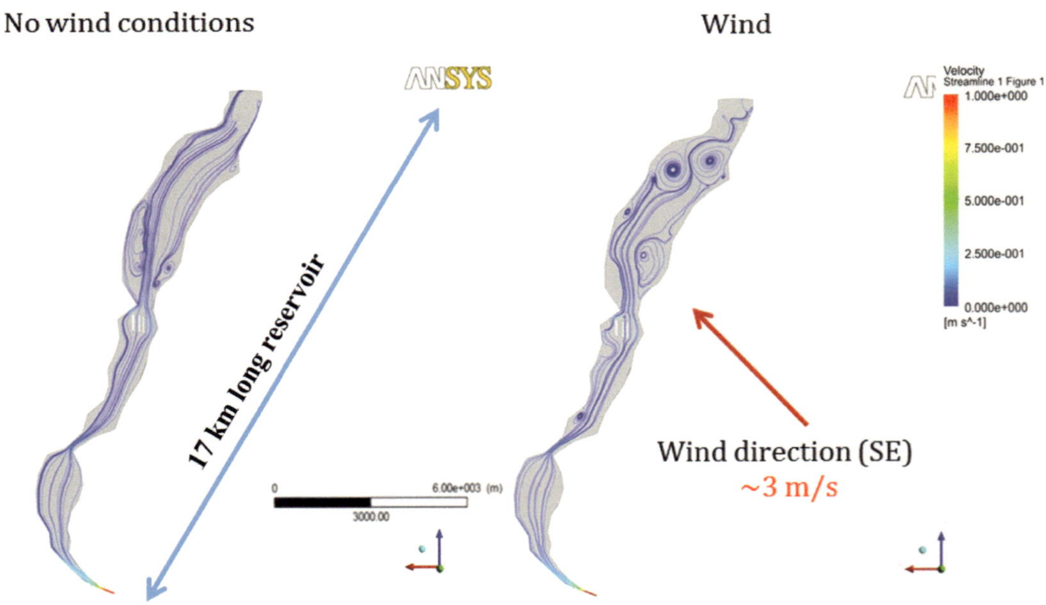

Figure 13. Comparison of velocity field (m/s) in the Sulejow reservoir in October at wind and no-wind conditions.

The findings suggest that when steady flow pattern develops in the basin, large regions of recirculation are formed below the outlet of the reservoir. **Figure 14** shows contours of velocity field in the Sulejow reservoir in (a) March (b) July, and (c) December.

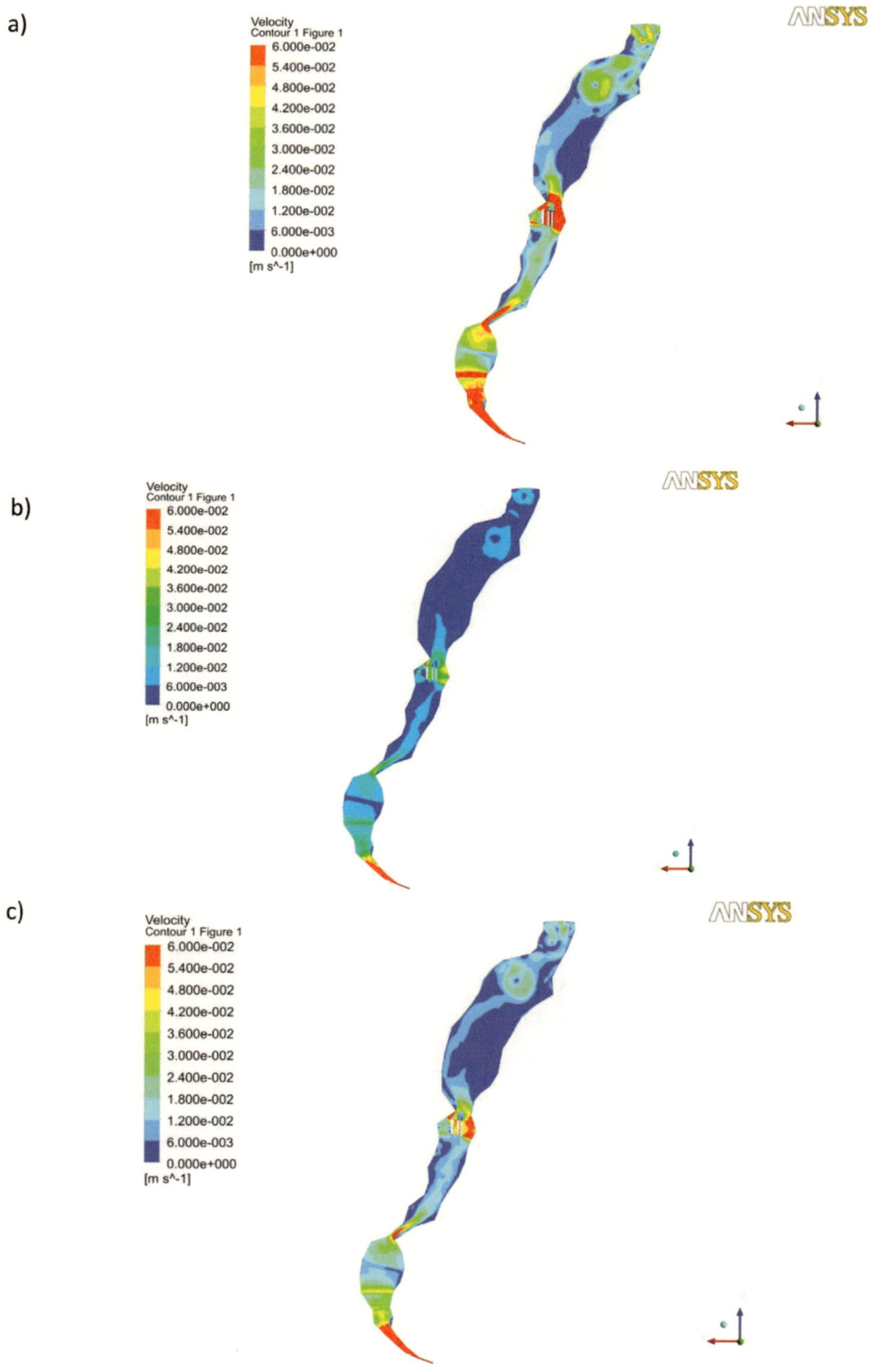

Figure 14. Contours of velocity field (m/s) in the Sulejow reservoir in (a) march (b) July, and (c) December.

5. Model verification

Numerical simulations have many advantages such as providing results in the entire domain and the ability to make changes to the geometry, boundary, or initial conditions, but numerical models always require validation of the simulations with reliable and appropriate experimental data [22]. The following section describes the measurements for obtaining data necessary for validation of the numerical simulations.

5.1. Acoustic measurements

Field measurements of velocity and turbulent quantities have been made much more accessible with the development of acoustic measurement methods. Acoustic Doppler Current Profiler (ADCP), highly efficient and reliable instrument for flow measurements in riverine and open-channel environments, has been used for the first time to determine the velocity profiles in the dam reservoir in Poland.

The measurements are repeated continuously during the movement of the boat. As a result, for a single passage along the cross section, a few hundred to several thousands of partial flow are obtained, which are summed during the measurement process. The measurements are repeated several times. The final result is calculated as the average value of at least four correct runs.

Teledyne RD Instruments' StreamPro Acoustic Doppler Current Profiler (ADCP) was used to validate the hydrodynamic CFD model. The device was mounted on a boat that moves across a transect of the reservoir channel. This technique can be adopted while complying the assumptions: (1) the water surface is not wavy and (2) velocity of the water is less than 2 m/s.

In order to correctly determine the places selected for the verification, the GPS Garmin eTrex 10 has been used. Flow velocity measurements were made in June 2013 and included four cross sections in the Sulejow reservoir and measurements of the flow rate at the inlet to the reservoir of the Pilica and Luciaza rivers.

The places, which have been selected for the verification of the hydrodynamic model, were arranged along the longitudinal axis of the Sulejow reservoir. The selection of these areas was dictated by expecting a different character of flow in indicated areas (**Figure 15**):

- Barkowice Mokre (1)—the place closest to the backwaters of the reservoir, located in the riverine zone, characterized by small depths (<3 m) and the highest flow rates.
- Zarzecin (2)—located in the upper, narrow part of the reservoir, with higher flow velocities, resulting in a half-river character. The depth at this point was about 4 m.
- Bronislawow (3)—situated in the central part of the reservoir, near the former water intake for Lodz City. The place is characterized by a low flow, due to greater width (1500 m) and depth of the basin (approximately 6 m).

- Tresta (4)—located in the lower part of the artificial lake, where the depth is about 7–8 m. The place is closest to the dam of the reservoir, characterized by, in addition to great depths, the largest cross-sectional width of approximately 2000 m.

5.2. Results of model validation

A reasonable agreement between the flow pattern predicted by the model and those deduced from the field data was found (**Figure 16**). To quantify the comparison, relative error was calculated by taking the difference between numerical and measured values and then dividing the results by the measured values. The measured velocity profiles were provided with a relative accuracy within the range 1–10%.

The possible reasons for the discrepancies are (1) inaccuracies in location of measuring points, (2) point velocity measurement errors, (3) errors in modeling the flow, and (4) errors in modeling the geometry.

The first category is related to the field velocity measurements taken from a boat. Considering the fact that a boat cannot maintain an absolute fixed position due to the waving and wind, errors are introduced in velocity measurements. A deviation of ±20 cm from the fixed position can cause large errors if there is a steep velocity change in the plane of measurements. The magnitude of this error could not be estimated accurately; however, rough estimate of the nearby velocities within a distance of ±20 cm at the measuring point resulted in an error in the range 3–5%.

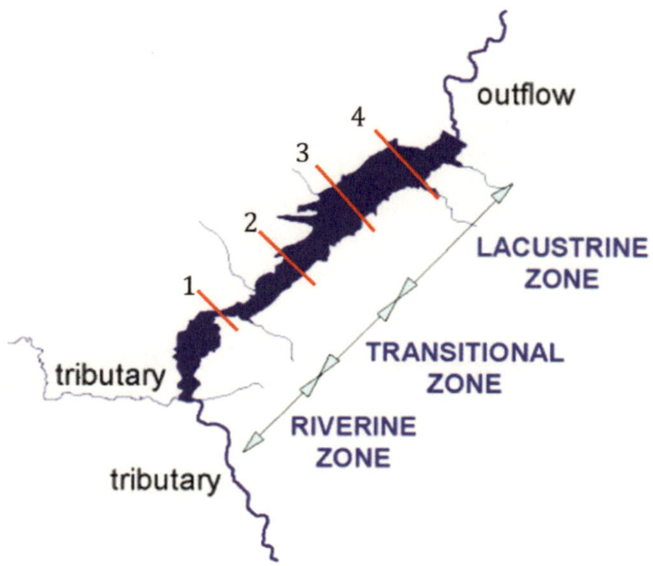

Figure 15. Hydrological and morphological differentiation of the Sulejow reservoir along the longitudinal axis.

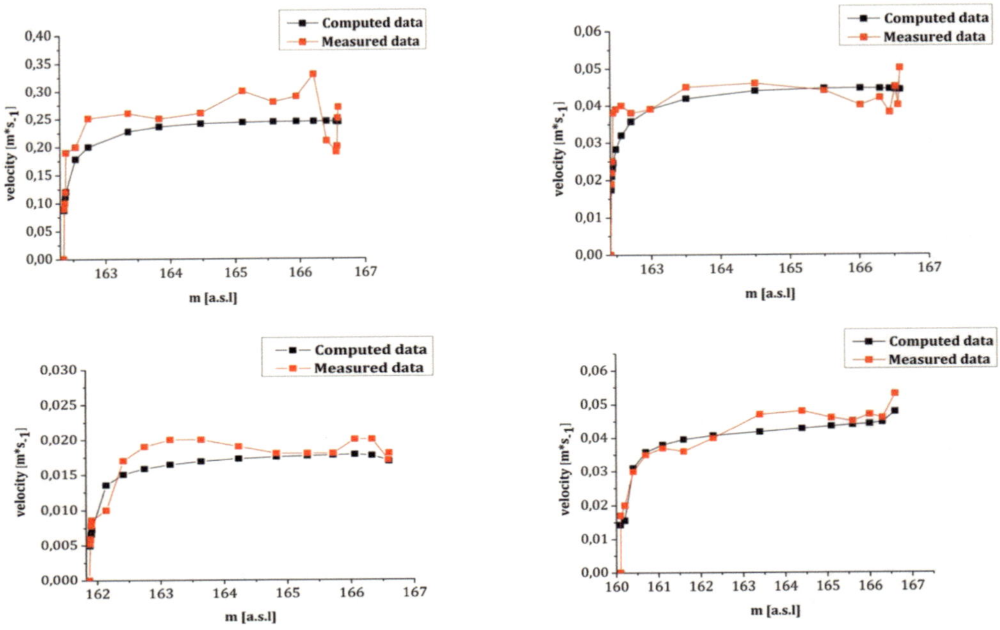

Figure 16. Comparison between computed and measured velocity profiles in four cross sections.

The second category consists of errors related to the instrument, its volume resolution, the range of operation, and the sampling time. The ADCP device could measure instantaneous 3-D velocity vectors with 1% accuracy. The vertical resolution of the instrument was 0.05 m, which is less than the vertical mesh spacing (Δy) used in the numerical model, that is, 0.08 m. In other words, the instrument resolution error can be ignored.

The third category is related to the numerical methods (discretization and iteration errors), the boundary conditions, and the closure models. For a carefully modeled problem that has well-posed boundary conditions, these errors are relatively low in comparison with other errors.

The fourth category is how the model geometry was built. The modeled geometry was an approximation of the reservoir topography as it was based on measurements of discrete cross sections. The regions between the cross sections were interpolated and may not represent the right topography of the artificial lake. A rapid variation in the topography significantly affects the flow velocity distributions. The spacing used in the present study was selected with special attention to the section properties of the reservoir. However, they may not have captured important changes of the bed.

Based on the above discussion, a total error between computed and measured velocities of about 10% is a reasonable assumption. Proper agreement of theoretical and experimental results shows the correctness of the actions, developed in the frame of this work.

6. Conclusions

The objective of this study was to develop and validate a three-dimensional numerical model for simulating flow through the long dam reservoir of a complex bathymetry (17 km length). As a result of the study, a three-dimensional one-phase CFD model of flow hydrodynamics in the large water body on the example of the Sulejow reservoir was developed with an accurate depiction of basin bathymetry and verified on the basis of field measurements.

The results of three-dimensional one-phase CFD model indicate that the flow field in the Sulejow reservoir is transient in nature, with visible swirl flows in the lower part of the lake.

The results of simulations confirm the pronounced effect of wind on the water flow in the reservoir and the accumulation of phytoplankton cells in the epilimnion layer of the lacustrine part of the Sulejow reservoir. Methodology developed in the frame of this work can be applied to all types of storage reservoir configurations, characteristics, and hydrodynamic conditions. Results of the simulation are complementary to the direct measurements of the surface water quality. A well-defined and constructed model can be used while developing a strategy for water environment quality control and can be used as an auxiliary tool for the monitoring and prediction of surface water quality and decision-making in the field of planning.

Author details

Aleksandra Ziemińska-Stolarska

Address all correspondence to: aleksandra.zieminska-stolarska@p.lodz.pl

Faculty of Process and Environmental Engineering, Lodz University of Technology, Łódź, Poland

References

[1] Hutter K. Hydrodynamics of Lakes. NY, USA: Springer-Verlag; 1984

[2] Imberger J, Hamblin PF. Dynamics of lakes, reservoirs and cooling ponds. Annual Review of Fluid Mechanics. 1982;**14**:153-187

[3] Serruya S, Hollan E, Bitsch B. Steady winter circulation in lakes constance and kinneret driven by wind and main tributaries. Archiv für Hydrobiologie. 1984;**70**(1):33-110

[4] Csanady GT. Large-scale motion in the great lakes. Journal of Geophysical Research. 1975; **72**:4151-4162

[5] Dubnyak S, Timchenko V. Ecological role of hydrodynamic processes in the Dnieper reservoirs. Ecological Engineering. 2000;**16**(1):181-188

[6] Fernando HJS. Turbulent mixing in stratified fluid. Annual Review of Fluid Mechanics. 1991;**23**:455-494

[7] Wüest A, Lorke A. Small-scale hydrodynamics in lakes. Annual Review of Fluid Mechanics. 2003;**35**:373-412

[8] Fischer HB, List EJ, Koh RCY, et al. Mixing in Inland and Coastal Waters. London: Academic Press; 1979. p. 483

[9] Imberger J. Transport processes in lakes: A review. In: Margalef R, editor. Limnology Now. A Paradigm of Planetary Problems. Amsterdam: Elsevier; 1994. pp. 99-193

[10] Lane SN, Richards KS. High resolution, two-dimensional spatial modeling of flow processes in a multi-thread channel. Hydrological Processes. 1998;**12**:1279-1298

[11] Sinha SK, Sotiropoulos F, Odgaard AJ. Three-dimensional numerical model for flow through natural rivers. Journal of Hydraulic Engineering. 1998;**124**(1):13-24

[12] Nicholas AP, Sambrook-Smith GH. Numerical simulation of three-dimensional flow hydraulics in a braided river. Hydrological Processes. 1999;**13**(6):913-929

[13] Dargahi B. Three-dimensional flow modeling and sediment transport in the river Klaralven. Earth Surface Processes and Landforms. 2004;**29**(7):821-852

[14] Fischer-Antze T. Assessing river bed changes by morphological and numerical analysis [PhD thesis]. Vienna Univ. Technology; 2005. p. 179

[15] Kennedy MG, Ahlfeld DP, Schmidt DP, Tobiason JE. Three-dimensional modeling for estimation of hydraulic retention time in a reservoir. Journal of Environmental Engineering. 2006;**132**(9):976-984

[16] Demuren AO. A numerical model for flow in meandering channels with natural bed topography. Water Resources Research. 1993;**29**(4):1269-1277

[17] Meselhe EA, Sotiropoulos F, Patel VC. Three-dimensional numerical model for open-channels. In: Proceedings of the ASCE Waterpower '95, San Francisco, CA, USA. Vol. 3. 1995. pp. 2315-2324

[18] Almquist CW, Holley ER. Transverse mixing in meandering laboratory channels with rectangular and naturally varying cross-sections. Technical Report CRWR-205. University of Texas; 1985

[19] Khosronejad A, Rennie C. Three-dimensional numerical modeling of unconfined and confined wall-jet flow with two different turbulence models. Canadian Journal of Civil Engineering. 2010;**37**(4):576-587

[20] Lane SN. Hydraulic modeling in geomorphology and hydrology: A review of high resolution approaches. Hydrological Processes. 1998;**12**(8):1131-1150

[21] Shepherd JF, Johnson CR. Hexahedral mesh generation constraints. Engineering with Computers. 2008;**24**(3):195-213

[22] Anderson A, Andreasson P, Lundström ST. CFD-modelling and validation of free surface flow during spilling of reservoir in down-scale model. Engineering Applications of Computational Fluid Mechanics. 2013;**7**:159-167

Chapter 4

Large-Scale Modeling of Dam Break Induced Flows

Amanbek Jainakov, Abdikerim Kurbanaliev and
Maralbek Oskonbaev

Additional information is available at the end of the chapter

http://dx.doi.org/10.5772/intechopen.78648

Abstract

Nowadays, mathematical modeling of the dam break flooding flows has become very interesting and challenging problem for scientists and engineers. In order to analyze possible risks and safety, mapping of flooded areas corresponding to different scenarios and conditions for the destruction of dams and reservoirs is significant. The results of mathematical modeling of large-scale flows in areas with a complex topographic relief were presented in this chapter. The flow was numerically simulated by the volume of fluid (VOF) method-based solver interFoam of OpenFOAM package, which solving the Reynolds-averaged Navier-Stokes equations (RANS) with the k-ε turbulence model. The mathematical model adequacy is checked by comparing with experimental data. The efficiency of the applied technology is illustrated by the example of modeling the breaking of the dams of the Andijan (Uzbekistan) and Papan (near the Osh town, Kyrgyzstan) reservoirs.

Keywords: dam break flows, topographic data, RANS, k-ε turbulence model, free surface, VOF, OpenFOAM

1. Introduction

The mathematical modeling using the advanced applied packages of computational fluid dynamics is an efficient tool for predicting various man-made and natural phenomena. The water flood of the regions and settlements takes the first place among natural spontaneous cataclysms in terms of the repetitions, extension area and annual material damage.

In this chapter, the problem of the prediction of consequences of a large-scale man-made disaster caused by the dam break is posed by the example of the Papan (near the Osh town, Kyrgyzstan) and Andijan (Uzbekistan) reservoirs.

IntechOpen

© 2018 The Author(s). Licensee IntechOpen. This chapter is distributed under the terms of the Creative Commons Attribution License (http://creativecommons.org/licenses/by/3.0), which permits unrestricted use, distribution, and reproduction in any medium, provided the original work is properly cited.

The adopted mathematical model is presented in Section 2. The general technology of the numerical solution of the adopted mathematical model, the initial and boundary conditions, the model discretization methods are presented in Section 3. Section 4 contains the results of verifying the package OpenFOAM by various dam break test cases. Finally, the examples of using the OpenFOAM package for modeling the breaks of the dams of the Papan and Andijan reservoirs are considered in Section 5.

2. The mathematical model of the dam break flows

The mass and momentum conservation laws for viscous incompressible liquid in the absence of mass forces lead to the following unsteady Navier-Stokes equations [1]:

$$\partial(\rho \overline{u}_i)/\partial x_i = 0, \tag{1}$$

$$\frac{\partial}{\partial t}(\rho \overline{u}_i) + \frac{\partial}{\partial x_j}\left(\rho \overline{u}_i \overline{u}_j + \rho \overline{u_i' u_j'}\right) = -\frac{\partial \overline{p}}{\partial x_i} + \frac{\partial \overline{\tau}_{ij}}{\partial x_j}, \tag{2}$$

where $\overline{u}_i$ are the mean velocity components, ρ is the density, $\overline{p}$ is the mean pressure, $\overline{\tau}_{ij} = \mu \left(\frac{\partial \overline{u}_i}{\partial x_j} + \frac{\partial \overline{u}_j}{\partial x_i}\right)$ is the mean tensor of viscous stresses, μ is the dynamic viscosity. The averaging is done in time, and the prime denotes the fluctuation part of the velocity. In the presence of external forces, it is necessary to augment these equations by the corresponding terms.

To close the systems of Eqs. (1) and (2), it is necessary to introduce some turbulence model. Most turbulence models employed in practice are based on the concepts of turbulent viscosity and turbulent diffusion. For the flows of general form, the turbulent viscosity introduced by Boussinesq, which couples the Reynolds stresses with the mean flow gradients, may be written in the following form [1]:

$$-\rho \overline{u_i' u_j'} = \mu_t \left(\frac{\partial \overline{u}_i}{\partial x_j} + \frac{\partial \overline{u}_j}{\partial x_i}\right) - \frac{2}{3}\rho \delta_{ij} k. \tag{3}$$

The turbulence kinetic energy k and its dissipation rate ε are determined from the following transport equations [1]:

$$\frac{\partial(\rho k)}{\partial t} + \frac{\partial(\rho \overline{u}_j k)}{\partial x_j} = \frac{\partial}{\partial x_j}\left(\mu + \frac{\mu_t}{\sigma_k}\right)\frac{\partial k}{\partial x_j} + P_k - \rho\varepsilon, \tag{4}$$

$$\frac{\partial(\rho \varepsilon)}{\partial t} + \frac{\partial(\rho \overline{u}_j \varepsilon)}{\partial x_j} = C_{\varepsilon 1} P_k \frac{\varepsilon}{k} - \rho C_{\varepsilon 2} \frac{\varepsilon^2}{k} + \frac{\partial}{\partial x_j}\left(\frac{\mu_t}{\sigma_\varepsilon}\right)\frac{\partial \varepsilon}{\partial x_j}, \tag{5}$$

where $P_k = \mu_t \left(\frac{\partial \overline{u}_i}{\partial x_j} + \frac{\partial \overline{u}_j}{\partial x_i}\right)\frac{\partial \overline{u}_i}{\partial x_j}$ is the rate of the generation of the turbulence kinetic energy by mean flow, and $\mu_t = \rho C_\mu \frac{k^2}{\varepsilon}$ is the turbulent viscosity.

The coefficients of the model have the following standard values [1]: $C_\mu = 0.09, C_{\varepsilon 1} = 1.44, C_{\varepsilon 2} = 1.92,$ $\sigma_k = 1.0,\ \ \sigma_k = 1.3.$

The method of determining the interface between two phases—water and air occupies special place at the modeling of the flow class under consideration. According to the main idea of the volume of fluid (VOF) method [1], one determines for each computational cell a scalar quantity, which represents the degree of filling the cell with one phase, for example, water. If this quantity is equal to 0, it is empty; if it is equal to 1, then it is filled completely. If its value lies between 0 and 1, then one can say, respectively, that this cell contains the free (interphase) boundary. In other words, the volume fraction of water α is determined as the ratio of the water volume in the cell to the total volume of the given cell. The quantity $1 - \alpha$ represents, respectively, the volume fraction of the second phase—air in the given cell. At the initial moment of time, one specifies the distribution of the field of this quantity, and its further temporal and spatial evolutions are computed from the following transport equation [1]:

$$\frac{\partial \alpha}{\partial t} + \frac{\partial(\alpha \overline{u}_i)}{\partial x_i} = 0. \qquad (6)$$

The free boundary location is determined by the equation $\alpha(x, y, z, t) = 0$. Therefore, the physical properties of the gas-liquid mixture are determined by averaging with the corresponding weight coefficient:

$$\rho = \alpha \rho_1 + (1 - \alpha)\, \rho_2, \mu = \alpha \mu_1 + (1 - \alpha)\, \mu_2. \qquad (7)$$

Here, the subscripts 1 and 2 refer to the liquid and gaseous phases.

The essence of the VOF method implemented in the solver interFoam of the OpenFOAM package [2] lies in the fact that the interface between two phases is not computed explicitly, but is determined, to some extent, as a property of the field of the water volume fraction. Since the volume fraction values are between 0 and 1, the interphase boundary is not determined accurately; however, it occupies some region, where a sharp interphase boundary must exist in the proximity.

3. Modeling technology

3.1. Initial conditions

In the case of unsteady problem, it is necessary to specify for the initial values for all dependent variables. The values of all velocity components are equal to zero because according to the condition of the problem under study, there is no motion until the moment of time $t = 0$. The hydrodynamic pressure is also equal to zero since the used solver—interFoam calculates hydrodynamic pressure [2]. The turbulence kinetic energy and its dissipation rate have some small value, which ensures a good convergence of the numerical solution at the first integration steps. The initial distribution of the volume fraction α is nonuniform because not all the computational cells are filled with water.

3.2. Boundary conditions

The no-slip condition is specified at solid walls of the computational region, which gives the zero components of the velocity vector. The Neumann conditions are specified for the water volume fraction: and $\partial\alpha/\partial n = 0$; at all wall boundaries, the fixedFluxPressure boundary condition is applied to the pressure (hydrodynamic pressure) field, which adjusts the pressure gradient so that the boundary flux matches the velocity boundary condition for solvers that include body forces such as gravity and surface tension [2].

The boundary conditions for the turbulence kinetic energy k and its dissipation rate ε were specified with the aid of the technique of wall functions [1]. Systematic calculations performed in this work show that the minimum value of dimensionless distance y^+ for all solid wall greater than 25, so we can use wall functions technique.

The influence of surface tension forces between the solid wall and the gas-liquid mixture were not taken into account.

The top boundary is free to the atmosphere so needs to permit both outflow and inflow according to the internal flow. That is why it is necessary to use a combination of boundary conditions for pressure and velocity that does this while maintaining stability.

3.3. Mesh generation and discretization of governing equations

BlockMesh utility was used for generation of mesh. The discretization of the computational domain is obtained by the control volume method [3]. The use of an upwind difference scheme for the convective and Gauss linear scheme for diffusion terms yields an acceptable accuracy of numerical computations.

The explicit Euler first-order method was used for the discretization of the unsteady term. Numerical solution of the unsteady equations coupled with the pressure was based on the PIMPLE method [2] with the number of correctors equal to 3.

The iterative solvers PCG and PBiCG—the methods of conjugate gradients and biconjugate gradients with preconditioning were used for solving the obtained system of linear algebraic equations. The procedures based on a simplified Cholesky's incomplete factorization scheme DIC and on the simplified incomplete LU factorization DILU were used as preconditioners.

Mesh sensitivity was analyzed for four mesh of 60 × 25 × 25, 90 × 40 × 40, 135 × 60 × 60 and 150 × 80 × 80, indicating that the mesh size was important only in the vicinity of the leading front location. Since of unimportant differences between the wave fronts using mesh 135 × 60 × 60 and mesh 150 × 80 × 80, a first mesh was adopted to reduce computational effort. In this case, the minimum value of dimensionless distance y^+ was more than 15 for all coordinate axes.

More detailed information about the boundary and initial conditions, discretization techniques, and the solution of systems of algebraic equations may be found in the work [2].

4. The OpenFOAM package verification

The problem on the liquid column collapse in a horizontal duct of rectangular cross section [4] is the first test problem. At the initial moment of time, the rectangular column of a viscous incompressible fluid is at rest. The column starts collapsing under the gravity force.

The numerical results (solid line) are compared with experimental data (markers) of the work [4] in **Figure 1**, where $a = 0.05715$ m is the water column width in the x-axis direction, $g = 9.81$ m/s^2 is the gravitational acceleration.

A good agreement between experiment data and results of numerical computation shows the reliability and accuracy of the present numerical modeling.

Figure 2 shows the configuration of the next model problem. The model represents a reservoir of hexahedral shape whose length is 3.22 m, height 2 m and width 1 m [5].

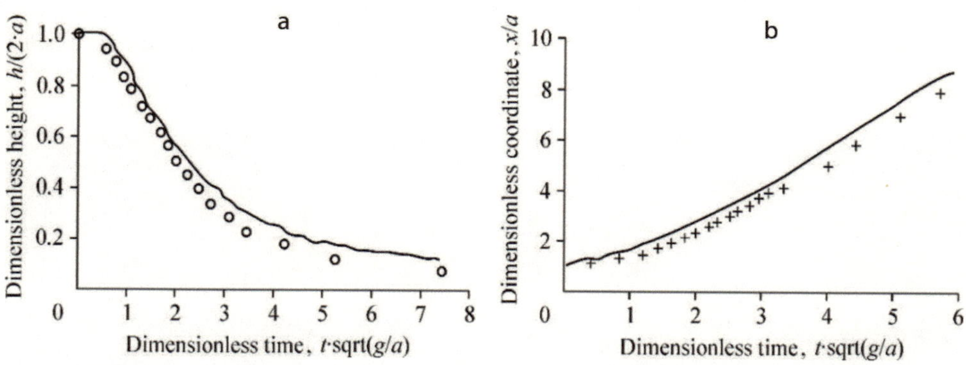

Figure 1. Comparison of numerical and experimental data. a—water column height change, b—the flow leading front location.

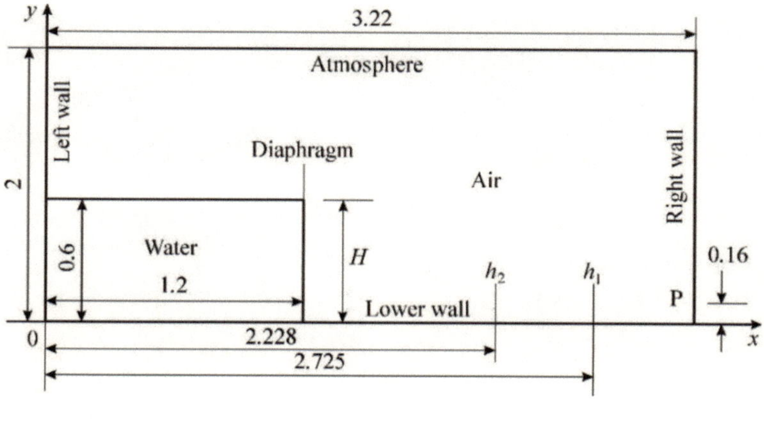

Figure 2. Test problem configuration.

The water column of height H = 0.6 m, length 1.2 m and width 1 m lies in the left lower corner of the reservoir. To measure the pressure force of fluid flow on the right reservoir wall, the corresponding pressure probe was placed at point P with coordinates x = 3.22 m and y = 0.16 m. Besides, the levels h_1 and h_2 of the water free surface were measured in two sections at x_1 = 2.725 m and x_2 = 2.228 m. The density of water was equal to 998.2 kg/m^3, and that of air was equal to 1.225 kg/m^3.

As initial configuration of the modeling with OpenFOAM, the water column in the left bottom part of the domain is at rest. When diaphragm is removed suddenly simulation is started, due to gravity water column drops and starts to move into the empty part of the domain. The total duration of modeling in time amounted to 2.5 s. In **Figure 3**, we have several snapshots of some stages of the simulation.

The diaphragm is suddenly removed at the moment of time t = 0, and the water column runs under the gravity force into the right empty part of the reservoir. At the moment of time $t \approx$ 0.65 s, the water reaches the right wall and impinging on it under the inertia force, moves upward. The flow is thinned as it moves upward along the right wall and at the moment of time t = 1.3 s, when the gravity force exceeds the inertia force; there occurs a reverse flow of

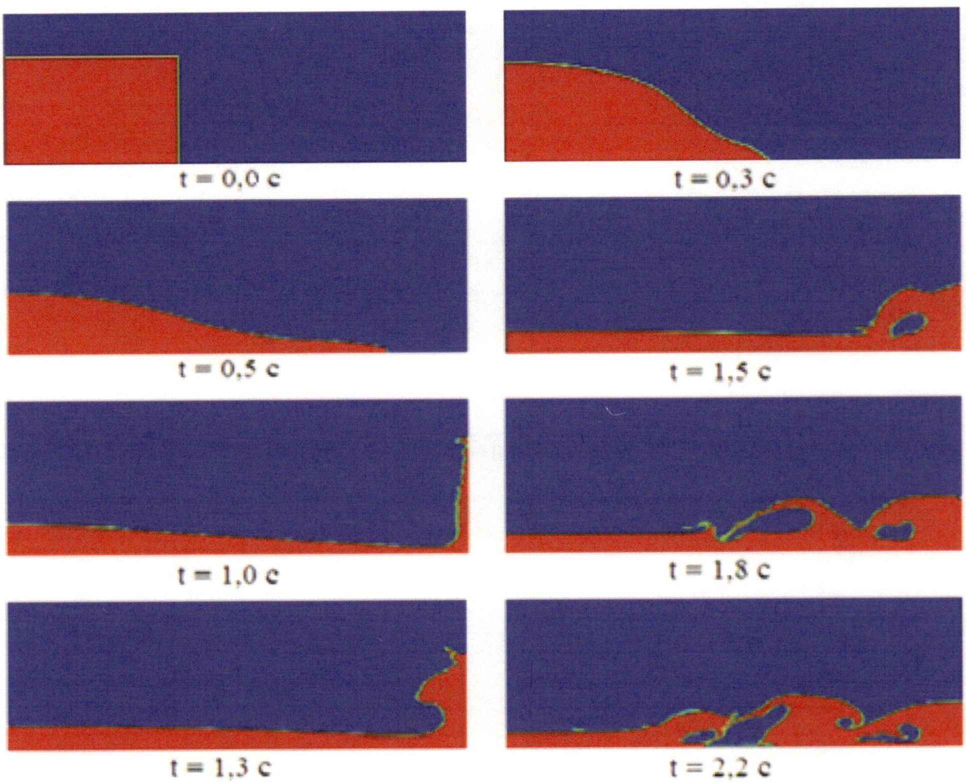

Figure 3. The motion of water column at different time step.

water with the formation of a typical bend of its surface. The reverse wave formed in such a way reaches the main flow, impinges onto it and forms the secondary wave, and so on (at $t = 1.5$ s and $t = 1.8$ s). After the moment of time $t = 2.2$ s, the water inertia drops significantly, and the further consideration of the motion is of no practical interest.

Figure 4 shows the numerical (solid line) and corresponding experimental (markers) data [5] on the water column height in sections with coordinates $x_1 = 2.725$ m and $x_2 = 2.228$ m. The coincidence between these data is satisfactory up to the moment of time $t \approx 1.5$ s for section $x_1 = 2.725$ m (**Figure 4a**).

After this moment of time, the reverse wave moving oppositely to the main stream impinges onto the free surface, and this gives rise to certain inaccuracies both in numerical and experimental data. The inaccuracies of such a kind were also observed in the work [6], where the well-known commercial package FLUENT was used for numerical modeling. The results of this work are presented in **Figure 5** in a form similar to **Figure 4**. The water column height (h_1, h_2) has been nondimensionalized here by quantity H, and the time is presented in the dimensionless form $\tau = t(g/H)^{0.5}$, where t is the physical time, $g = 9.81$ m/s^2 is the free-fall acceleration, $H = 0.6$ m is the initial water column height. One can conclude from a comparison of **Figures 4** and **5** that the free surface levels in two different sections have been predicted more accurately in the present work.

The problem of determining the pressure P on the obstacle is very important at the solution of unsteady problems with free surface, in particular, at the interaction of forming waves with various obstacles. **Figure 6** shows the numerical results (solid line) for the pressure of the fluid on the right wall at point P with coordinates ($x = 3.2$ m, $y = 0.16$ m) and the corresponding

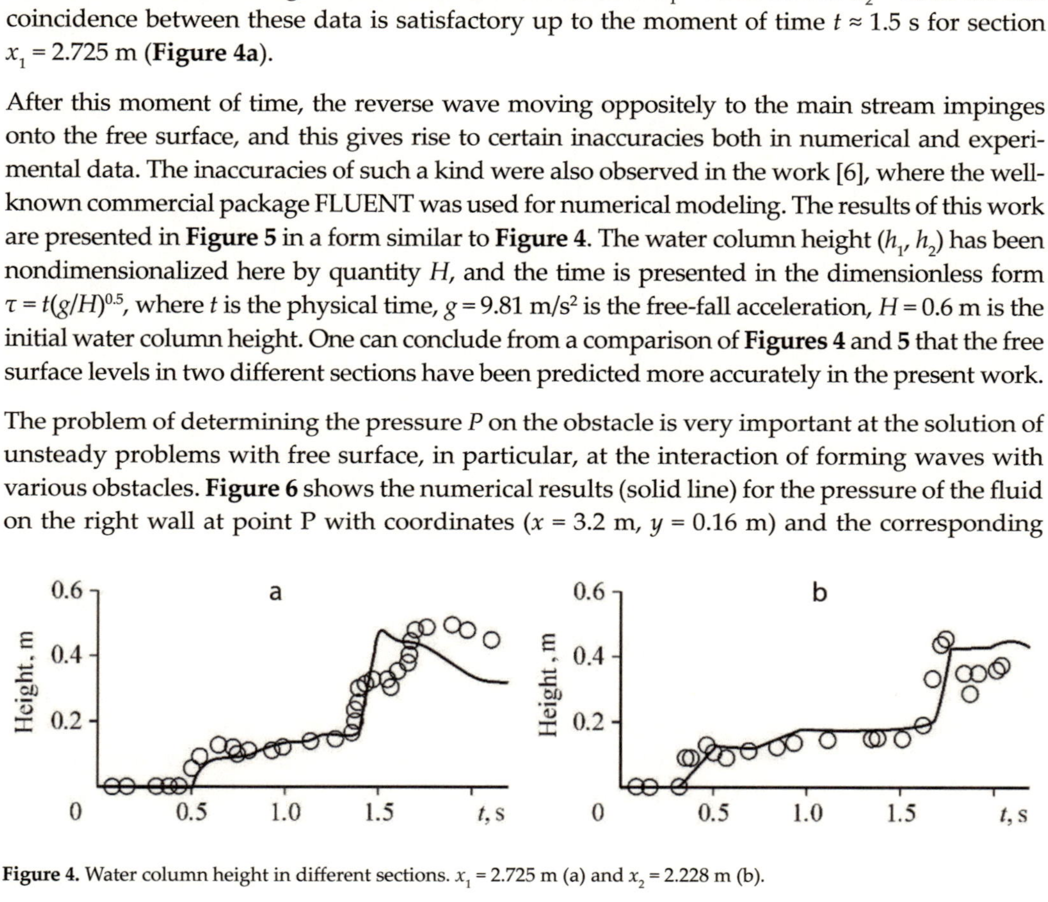

Figure 4. Water column height in different sections. $x_1 = 2.725$ m (a) and $x_2 = 2.228$ m (b).

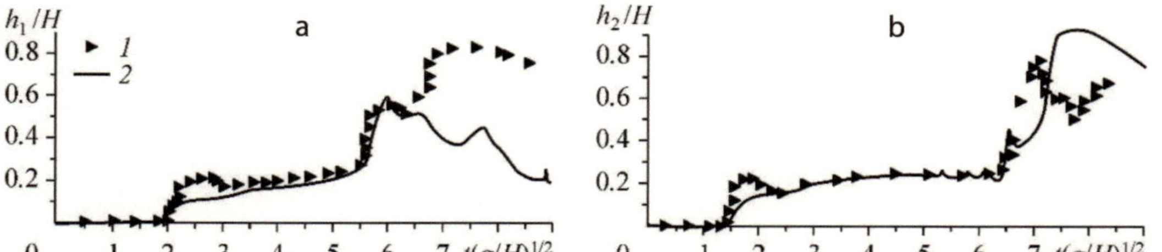

Figure 5. Water column height from the work [6] at $x = 2.725$ m (a) and 2.228 m (b). 1—experiment [5], 2—computation of the work [6].

experimental data (markers). The exact pressure value at point P cannot be measured because the pressure probes have a finite size—a circle about 90 mm in diameter.

The numerical pressure at the pressure probe center (see **Figure 6a**) increases slowly with time, after the moment of time t = 1.5 s or after the second maximum the coincidence of experimental data with numerical results improves. The character of the numerical pressure at the lower point of the probe (see **Figure 6b**) agrees fairly well with the character of the variation of corresponding experimental data; however, the maximum values are slightly underestimated.

It is assumed at the numerical modeling that the diaphragm is removed suddenly that is its vertical velocity is infinite. On the other hand, there can be also different physical conditions, which are hard to take into account at the numerical modeling. A detailed analysis of the conditions for conducting the experiment shows that this velocity has a finite value. The verification experimental data under the same conditions give different results, which do not coincide with one another [5]. In addition, the above discrepancies between the computation and experiment after the moment of time t = 1.5 s can probably be explained by the two-dimensionality of the employed model. It is possible that the flow acquires a three-dimensional character at some points of the computational region.

Comparing the data of the present work (**Figure 4**) and the work [6] (**Figure 5**), one can assert that the numerical results of modeling the task under consideration, which were obtained with the aid of the open package OpenFOAM, are closer to the experimental data than the results obtained with the aid of the commercial package FLUENT.

The problem of the fluid column breakdown in a reservoir of rectangular shape with an obstacle [7] is the next, more complex test problem. The chosen coordinate system and the problem diagram without the geometric proportion preservation are shown in **Figure 7**.

An open reservoir 3.22 m in length with cross section of 1 × 1 m^2 has been used in experiment. The reservoir was initially partitioned into two unequal parts by a vertical wall located in section x = 2 m. The water 0.55 m in height is located behind this wall, another reservoir part is empty. A container 40 cm in length with cross section of 16 × 16 cm^2 is located in this empty part of reservoir. The container left-face coordinate equals x = 0.67 m.

At the execution of experiment, the water column height and the fluid pressure on the container surface were measured. The location of measuring probes is shown in **Figure 8**. Four probes were used to measure the water column height: the one on the part filled with

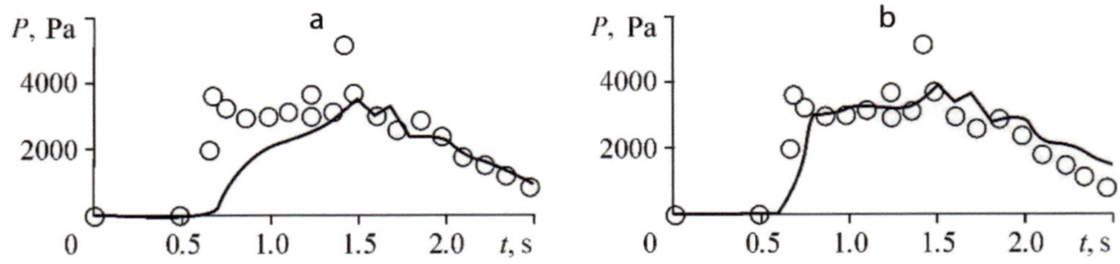

Figure 6. Pressure variation at point P. Pressure at the center (a) and at the lower point (b) of the probe.

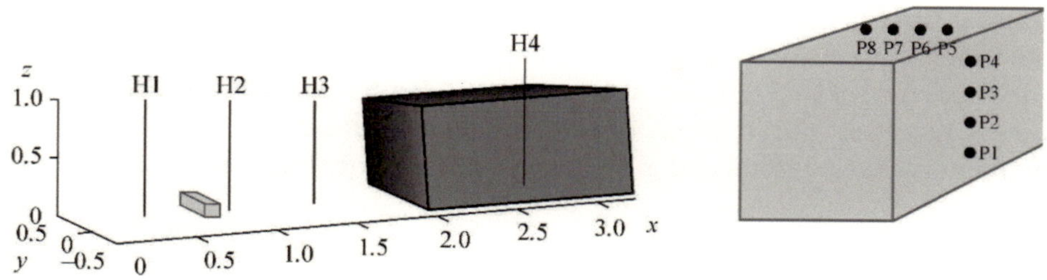

Figure 7. Configuration (on the left) and location of measurement pressure probes (on the right).

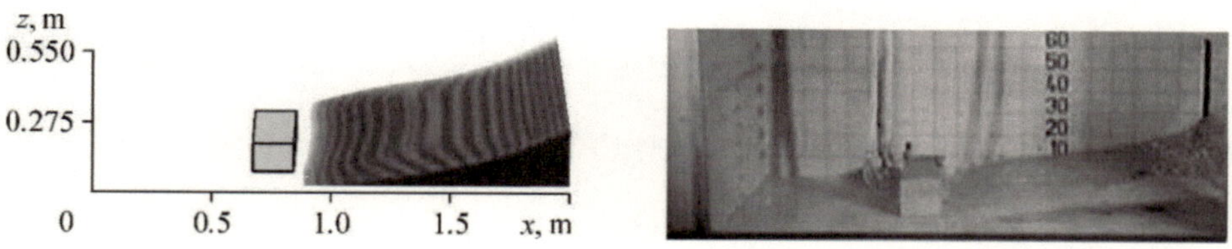

Figure 8. Comparison of numerical (on the left) and experimental data [7] (on the right) at the moment of time $t = 0.4$ s.

water—H4, and the remaining ones H1, H2, and H3 were located in the reservoir empty part. The coordinates of these probes are $x = 0.5$, 1.0, 1.5 and 2.66 m, respectively.

The container was supplied with eight pressure probes: four probes on the exterior surface at points with coordinates $z = 0.025$, 0.063, 0.099 and 0.136 m, and the remaining four probes were located on the container upper side with coordinates $x = 0.806$, 0.769, 0.733 and 0.696 m. The probes on the exterior surface were located at the distance of 0.026 m to the left of the central line ($y = 0$), and the probes on the upper surface were located at the distance of 0.026 m of this line. The force exerted on the container on the water stream side was also measured in experiment.

The water was at rest up to the moment of time $t = 0$. At the moment of time $t = 0$, the separating wall was removed suddenly, and the water column ran into the reservoir empty part under the gravity force. The 180 × 60 × 80 computational grid was used, and the CPU time amounted to 6 s. The initial time step was 0.001 s, and it was varied further depending on the Courant number, which was equal to 0.85.

Figures 8 and **9** show the comparison of numerical and experimental data for the moments of time $t = 0.4$ and 0.6 s, respectively. The pictures of shooting done during the experiment are shown on the right. One can notice a fairly good visual coincidence of the numerical results with experimental data.

The time of reaching the container by water flow both in experiment and at the numerical simulation is the same. Besides, the free surface shapes forming after the flow impact onto the container also coincide. One can note, however, that there are at the numerical modeling some imperfections of the free surface between water and the ambient medium—air.

Figure 10 shows the water flow height at two different points: in the reservoir and in the immediate proximity of the container. There is a fairly good agreement between them until the water returns from the back wall after the moment of time $t \approx 1.8$ s. After that, the numerical data (solid lines) prove to be somewhat higher than the experimental ones (markers). At the moment of time of $t \approx 5$ s, the secondary wave reaches the neighborhood of the probe H2. This time is, however, equal to about 5.3 s at the numerical modeling. The general character of the variations of the numerical and experimental data nevertheless coincides.

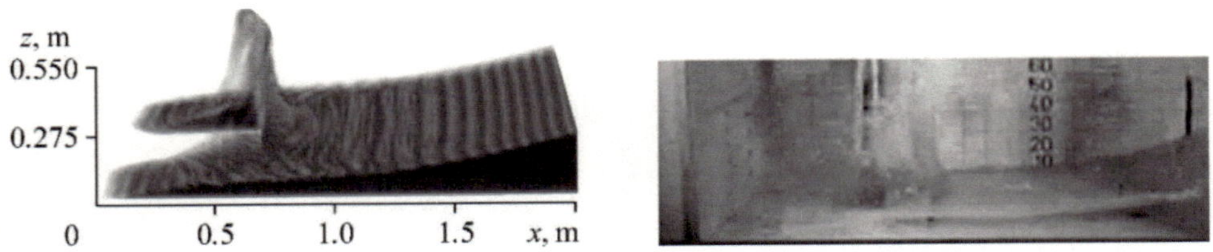

Figure 9. Comparison of numerical (left) and experimental [7] (right) data at the moment of time $t = 0.6$ s.

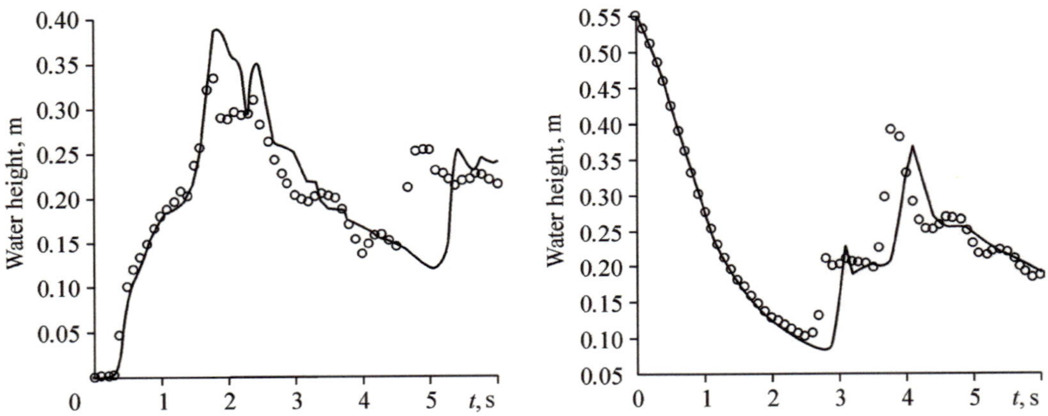

Figure 10. Water flow heights at points H2 (left) and H4 (right).

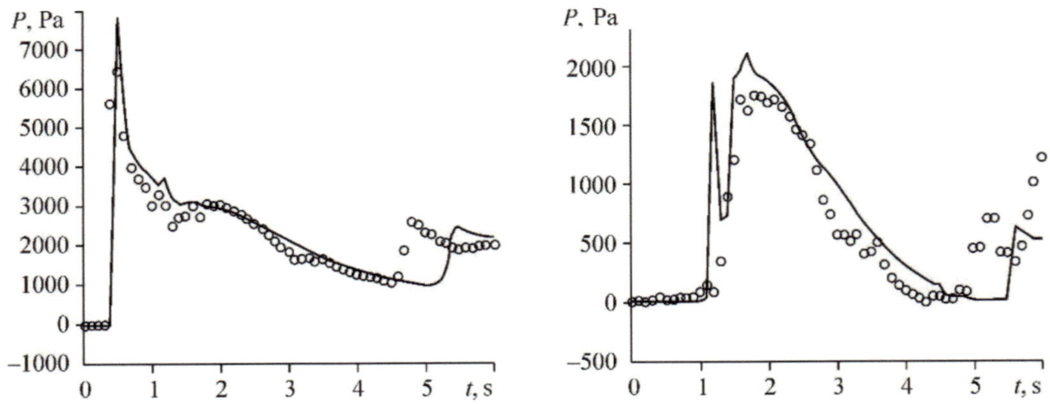

Figure 11. Pressure at points P2 (left) and P7 (right).

Figure 11 shows the moment of time $t = 0.5$ s when the wave reaches the container has been predicted with a good accuracy; however, the computed pressure value (solid lines) is slightly overestimated as compared to the experimental value (markers) (the left figure).

The numerical values of the second pressure maximum at point P2 are, however, shifted in comparison with experiment to the right by 0.6 s, and at point P7, they are shifted by 0.5 s. As the experiment has shown, the moment of time when the flow reaches repeatedly the container (≈4.7 s) is seen in these figures fairly well. Besides, when comparing the numerical and experimental pressure values at points P7 (the right figure) one can notice some differences. After $t = 1.3$ s, a small oscillation lasting for about 0.2 s takes place in numerical computations, which is not observed in experiment.

5. Dam break flooding flow modeling in real region

To illustrate the techniques of the application of numerical modeling of large-scale hydrodynamic computations, we consider the problem of computing the flood process in the areas near the dams of the Andijan and the Papan reservoirs (see **Figure 12**).

It is to be emphasized here that the situation of a real breakthrough of the dam and the flood of the areas at the lower level is not modeled here but the fundamental possibility of using the above technology under the availability of necessary topography data is demonstrated. The topography data of Digital Terrain Elevation Data [8] were used in computations, which were converted subsequently into the STL format. The hexahedral background grid generated with the aid of the utilities blockMesh and snappyHexMesh of the OpenFOAM package was transformed into a three-dimensional surface, which is employed for modeling the flood process (**Figure 13**).

For the Andijan reservoir, the computational field had the sizes 6000 × 4000 × 1500 m, the physical modeling time amounted to about 9 h for the 120 × 120 × 80 grid. **Figures 14** and **15** show different stages of the flood in areas with real topology. The red corresponds to a pure

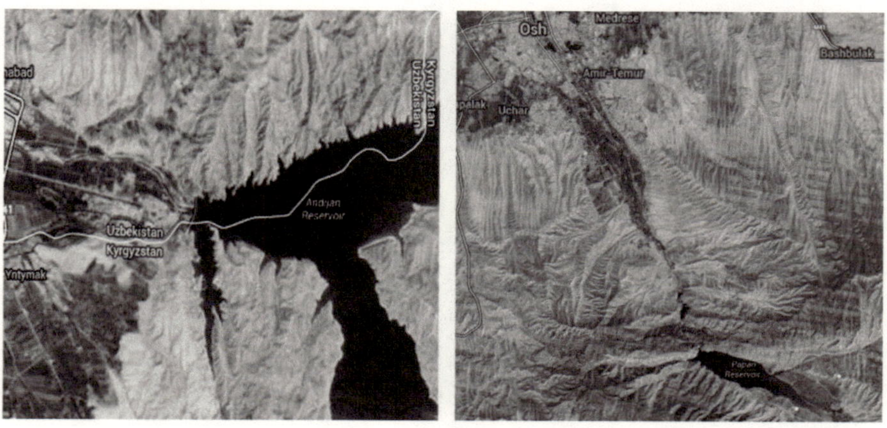

Figure 12. Maps of the Andijan (left) and Papan (right) reservoirs.

water flow, and the blue corresponds to air flow (there is no water flow in blue regions). It is seen in **Figure 14** that the leading front of water flow reaches during 240 s the lower boundary of the computational region passing a distance of about 6000 m, covers the most part of the area located downstream.

The computational field for the Papan reservoir has the sizes 5000 × 5000 × 1300 m (see **Figure 15**).

The total computing time in the case of the 50 × 60 × 30 grid amounts to about 5 h. As is seen in **Figure 15**, after the moment of time $t \approx 200$ s there forms a reverse flow (**Figure 15d**

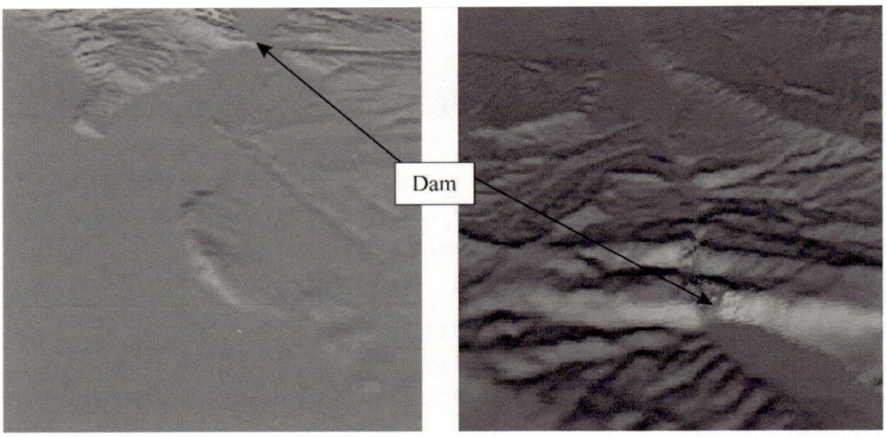

Figure 13. Three-dimensional surface of the Andijan (left) and Papan (right) reservoirs areas.

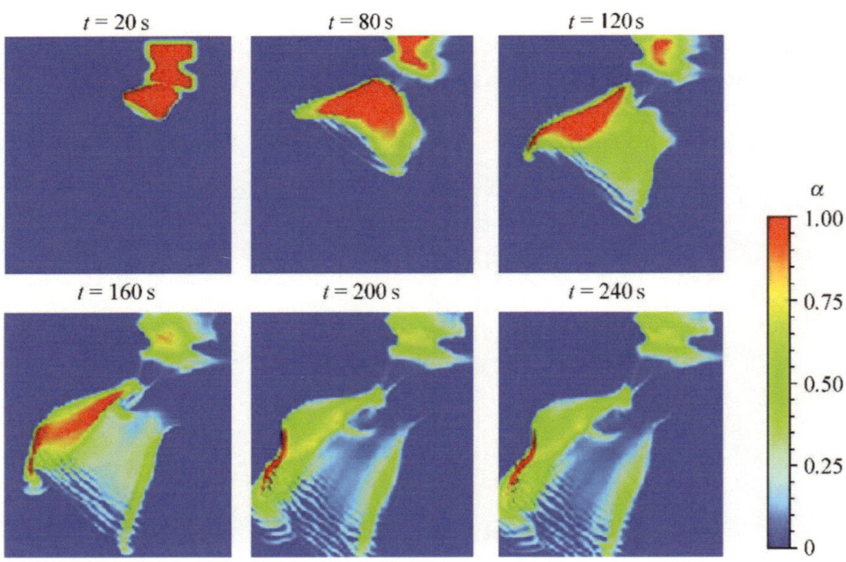

Figure 14. Flow pattern for the Andijan reservoir.

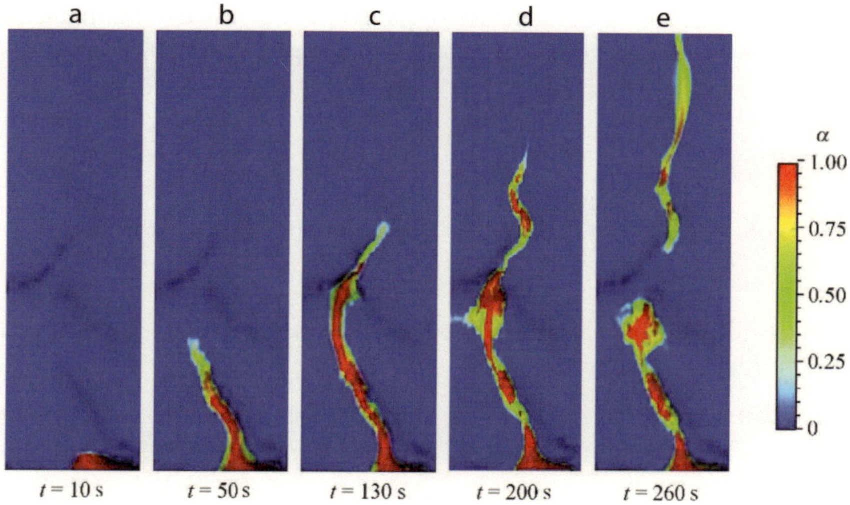

Figure 15. Flow pattern for the Papan reservoir.

and **c**) and after the moment of time t = 260 s; it separates into two parts—one part is in the zone of reverse flows and the other continues its flow in the lower part of the river bed. The computations show that about 60% of the entire initial water volume remains in the zone of reverse flows.

6. Conclusions

The results of the mathematical modeling of complex hydrodynamic phenomena on the basis of unsteady three-dimensional Navier-Stokes equations describing the dynamics of a gas-liquid mixture with free boundary have been presented. The adequacy of the employed model has been verified by the example of the classical problems of computational fluid dynamics. Special attention has been paid to the accuracy of the computation of the water flow level and the gas-liquid flow pressure on the reservoir walls. The efficiency of the employed technology has been illustrated by the example of modeling the breaks of the dams of the Andijan (Uzbekistan) and Papan (near the Osh town, Kyrgyzstan) reservoirs. The developed technology is universal and can be used for the flood modeling for a real relief. It is shown that the relief features are a substantial factor.

Acknowledgements

The Dam break flooding flow modeling in real region part of this chapter was done within the framework of the Fulbright exchange program of the US State Department. Authors gratefully thanks to of Prof. Greg Olyphant from the Indiana University, Bloomington, USA for his supervision and to Sally L. Letsinger, PhD, research Hydrogeologist of the Center for Geospatial Data Analysis of Indian Geological and Water Survey of Indiana University, Bloomington, USA for helping to prepare all necessary topographic data.

Author details

Amanbek Jainakov[1], Abdikerim Kurbanaliev[2*] and Maralbek Oskonbaev[2]

*Address all correspondence to: kurbanaliev@rambler.ru

1 Asanaliev Institute of Mining and Technologies of the Razzakov Kyrgyz State Technical University, Bishkek, Kyrgyzstan

2 Osh State University, Osh, Kyrgyzstan

References

[1] Ferziger JH, Peric M. Computational Methods for Fluid Dynamics. Berlin: Springer Verlag; 2002. DOI: 10.1007/978-3-642-56026-2

[2] OpenFOAM Foundation. 2018. Available from: https://openfoam.org/ [Accessed: 2018-03-30]

[3] Patankar SV. Numerical Heat Transfer and Fluid Flow. New York: Hemisphere Publ. Corp.; 1980. DOI: 10.1002/cite.330530323

[4] Martin JC, Moyce WJ. An experimental study of the collapse of liquid columns on a rigid horizontal plane. Philosophical Transactions of the Royal Society A: Mathematical, Physical and Engineering Sciences. 1952;**244**(882):312-324. DOI: 10.1098/rsta.1952.0006

[5] Zhou ZQ, Kat JOD, Buchner B. A nonlinear 3D approach to simulate green water dynamics on deck. In: 7th Intern. Conf. on Numerical Ship Hydrodynamics; Nantes, France. 1999. pp. 5.1-1-5.1-15

[6] Abdolmaleki K, Thiagarajan KP, Morris-Thomas MT. Simulation of the dam break problem and impact flows using a Navier-Stokes solver. In: 15th Australasian Fluid Mechanics Conf. The University of Sydney; Australia; 13–17 December. 2004. pp. 135-138

[7] Kleefsman KMT, Fekken G, Veldman AEP, Iwanowski B, Buchner B. A volume-of-fluid based simulation method for wave impact problems. Journal of Computational Physics. 2005;**206**(1):363-393. DOI: 10.1016/j.jcp.2004.12.007

[8] Digital Terrain Elevation Data (DTED). 2012. Available from: http://data.geocomm.com/catalog/KG/group121.html [Accessed: 2012-02-22]

Chapter 5

Influence of Geological Structure on Dam Behavior and Case Studies

Arash Barjasteh

Additional information is available at the end of the chapter

http://dx.doi.org/10.5772/intechopen.78742

Abstract

Complex engineering projects including large dams require extensive reconnaissance. The study of geological relationships is therefore of major importance, with emphasis on the characteristics of the geological structures. Accordingly, geologic structure affects dam site and reservoir behavior in three ways: (1) its impact on the geomechanical properties of rocks; (2) the importance of geologic structures in the identification and assessment of karst hydrogeology; and (3) its role in seismotectonic and seismic risk analysis of dam projects. Site geology and availability of various geologic data obtained from site investigation are key points in dam construction. Geological structure plays an important role in dam site geology and imposes major limitations on dam behavior during and after construction stages. This role has its own effect on major subjects such as: morphotectonics of rivers; geotechnical properties and engineering geology of dam sites; and hydrogeology of dam abutments and reservoir. The variability and complexity of geological structures regarding their tectonic situation result in different scenarios regarding dam's behavior. This chapter examines the link between geological structure and dam behavior during and after construction period by describing four dam case examples: two earth (Marun and Gotvand) and two concrete (Karun-1 and -3) dams in Iran.

Keywords: geologic structure, dam behavior, Karun-1 dam, Karun-3 dam, Marun dam, Gotvand dam, southwestern Iran

1. Introduction

Complexity of large civil engineering projects such as dams necessitates extensive reconnaissance. Particularly, the construction of dams and reservoirs in karstified rocks with known high perviousness due to dissolution faults and conduits intensifies the risk of water loss and

IntechOpen

© 2019 The Author(s). Licensee IntechOpen. This chapter is distributed under the terms of the Creative Commons Attribution License (http://creativecommons.org/licenses/by/3.0), which permits unrestricted use, distribution, and reproduction in any medium, provided the original work is properly cited.

instability [1]. Water seepage and loss through dam foundation and abutments, especially those constructed in karstic regions lead to considerable costs, sometimes postpones dam construction targets. Although this is of less importance regarding dam instability, current experience shows that there could be a fair relationship between instability and water seepage, where fault and fracture systems are the main conduits for water flow. Thus, the study of geological relationships is of major importance, with main emphasis on the characteristics of the structural features and recent tectonic movements [2]. The structural geology has a great deal in contributing to the engineering projects, and knowledge of structural geology setting is essential for safe design of these projects [3]. A detailed study with focus on the analysis of fault and fracture systems, their activities, movement types, and position ought to be done. The evaluation of stress relationships and possible temporal and spatial deformations are also necessary.

On a regional scale, structural relationships have an important role in forming the hydrogeological properties in karst regions. Although the possibility for karstification perpendicular to the geologic structures is considerably reduced, where an anticlinal structure exists between the reservoir and the lower erosion base level [4], in young orogens such as the Zagros belt, where transverse and antecedent drainage pattern is of particular importance [5], reservoir water tightness is highly risky due to dissolution fractures and conduits. One of the structural controls in the development of drainage in modern orogens is fault and fracture systems,

Figure 1. Locations of Karun-1 dam (1), Karun-3 dam (2), Marun dam (3), and Gotvand dam (4) considered for this study.

due to actively growing of structures [6]. In the Zagros fold belt, superimposed drainage was found to be related to structural associations leading to transverse drainage pattern [7]. The position of the folded structures had a decisive role in the formation of the pre-karstic drainage network, that is, directions of the main surface outflow. As a result of the new tectonic activity, the homogeneous anticlinal structure has been fractured transversely and huge separate blocks have been formed [8]. Accordingly, challenges regarding the relationship between geologic structure and dam behavior are described and discussed for four dam cases, namely Karun-1 concrete dam, Karun-3 concrete dam, Marun earth dam, and Gotvand earth dam, which are located in Khuzestan Province, southwest of Iran (**Figure 1**).

2. General geology and seismicity

The Zagros fold-thrust belt (briefly named as Zagros mountains) being located to the northeast of the Persian Gulf extends for about 1800 km between northern Iraq and the Strait of Hormuz and comprises the deformed part of the Arabian plate following its continental collision with central Iran almost since the Miocene [9]. It is a branch of the Alpine-Himalayan Orogenic belt [10, 11]. The mountain range can be divided into two distinct zones say, the northeastern High Zagros (**Figure 2**) and the southwestern Simply Folded Belt (SFB) based on topography, geomorphology, exposed stratigraphy, and seismicity [12]. The Simply Folded Belt extends from the High Zagros Fault to the Persian Gulf [13]. Topographic highs are typical of the anticlines and synclines form the topographic lows. The existing landforms graphically reveal the geologic structure. Anticlines show remarkable regularity in relief over long distances and the rivers which cut them usually do so at sharp angles to the strike of the anticlines. The SFB is a fold belt characterized by NW-SE trending parallel anticlines and synclines and is composed of elongated whaleback or box-shaped anticlinal mountains. From a geomorphological view point, the anticlines can be divided into two sets; plain anticlines and mountain anticlines [12].

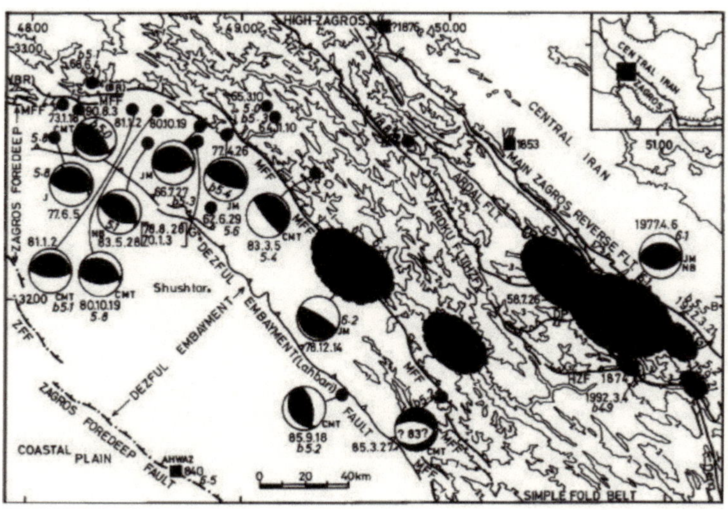

Figure 2. Seismotectonic features of the study area in the Khuzestan Province [14].

In fact, most part of the study area, that is, Khuzestan Province and the aforementioned dams are almost located in the Dezful Embayment, which is a structural unit of the SFB.

Iran can be generally divided into five major seismotectonic zones (**Figure 2**) that are subjected to destructive earthquakes excluding one (Central Iran characterized by low seismicity). It is considered as a broad seismic zone over 1000 km width that extends from the Turanian platform (southern Eurasia) in the northeast to the Arabian plate in the southwest. The Iranian plateau is characterized by active faulting, recent volcanic activity, and high density of active and recent faults. Reverse faulting dominates the tectonic mechanism of the region. Southwest of Iran, where the studied large dams are located, belongs to the Zagros Active Fold Belt from the seismotectonic point of view [14]. Seismicity in this belt correlates well with topographic elevations greater than 1.5 km. Fault plane solutions for several earthquakes consistently show high angle (40°–50°) reverse faulting and the estimated depths range from 8 to 13 km with magnitudes that range from 4 to 6. The rate of seismicity in this zone is higher than the others, but the type of seismicity is mostly between small to moderate and seldom large. Due to its particular tectonic condition, the large earthquakes have rarely been accompanied by surface rupture in the SFB. Based on the available fault plane mechanisms of the regional events, the maximum principal stress, which is due to regional tectonic forces, strikes N30°± 5° (NE–SW) [15].

3. Karun-1 dam

Karun-1 (also known as Shahid Abbaspour) is a 200 m high concrete arch dam (**Figures 1** and **3**) at nearly 52 km to the northeast of Masjid-I-Solaiman City in the Khuzestan Province. It was constructed on the Karun River with a reservoir capacity of 3139 MCM to produce electrical energy, to control flood, and to regulate water. The first impounding of the dam reservoir was in December 1976. The presence of two springs downstream the right abutment of the dam site and construction of a new power plant besides recent rock sliding on the dam abutments are the main points of discussion. Since the completion of the dam, seepage problem was a key challenging subject regarding the right abutment shear zone and downstream springs [16–18].

Figure 3. Upstream view of Karun-1 dam site before (left) and after (right) construction.

http://dx.doi.org/10.5772/intechopen.78742

Some studies were later done regarding the construction of a new second powerhouse in the left abutment of the dam [19–21] and recently, rock fall on the right abutment [22, 23].

Karun-1 dam site is located on the southwestern limb of Kamarun anticline (**Figure 4**) with average bedding dip of 35° toward SW. The anticline is composed of Asmari Limestone of Olig-Miocene age. Asmari formation limestone is a suitable rock foundation for dams regarding its relatively exclusive characteristics such as rigidity and morphology. This formation constitutes a series of double plunging, asymmetrical folds with northwest-southeast trend having steeper southern flanks than the northern ones. The Asmari limestone forms the entire foundation of the Karun-1 dam. It is divided into three parts at the dam site namely, lower, middle, and upper Asmari. The dam is situated on low-karstified middle Asmari that consists of a relatively permeable zone, which in turn is overlain by an impervious shale layer. The upper Asmari limestone exposed just downstream the dam, is highly karstified [16, 18]. The anticline shows some axial plane rotation along its southeast plunge, however, its general trend is northwest-southeast as is common in the whole belt.

The region is seismotectonically very active regarding its location in the Zagros active fold-thrust belt [14]. The main faults in the region are of thrust types of which Izeh fault zone cutting across east of the region is a very known feature due to its right lateral component of movement. The Andeka Fault is the main active fault close to the dam site that is characterized by very recent activity [24, 25]. Although no major fault exists at the dam site, recent investigations for excavation of a new powerhouse identified a fault with a general NW-SE trend [20], accompanied by high fracture density. This is most probably a hidden blind fault that is expected to cut the anticline core. The geologic structure of the dam site includes bedding, joint sets, and a spectacular shear zone in the right abutment. Recent study indicated a tectonic lineament with a NE-SW trend that passes through the right abutment, which could

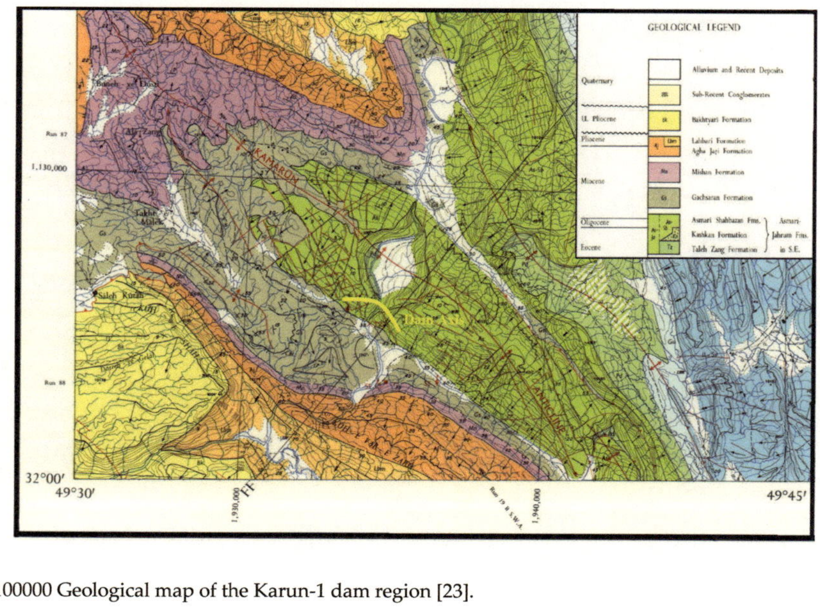

Figure 4. 1/100000 Geological map of the Karun-1 dam region [23].

be the main cause of shearing and fracturing in the right abutment [23]. The bedding plane has an average attitude of 32.5/210 (dip/dip direction) in the southwest limb (**Figure 5**) on which the dam is located and is expected to be the major discontinuity at the dam site. The southwest limb has a larger dip than the northeast one, which is characteristic of the southwestern limbs of the Zagros anticlines. It shows high degree of separation with less favorable shear signature. Three joint sets are identified at the dam site with 24/253, 21/166, and 47/038 attitudes. The joints are filled with calcite or clay and partly show slicken-lines. Their spacing is mostly dependent on strike and dip, but small changes can be seen in the joint opening of the left abutment after the excavation of the new powerhouse (No. 2 powerhouse) possibly due to blasting operation. Big or Sabz (meaning green) and Powerhouse Springs are the most significant and spectacular hydrogeological features at the dam site [18]. They demonstrate a widespread karst system in the Asmari limestone. The Zagros anticlines, particularly in the Asmari Formation, contain tension-induced, open fracturing, which has introduced significant secondary permeability to the rock. In this regard, the right abutment shear zone trend strikes along this direction that corresponds to the main regional extensional fractures and is very favorite for ground water flow. The Big Spring flows through a large karst cavern with an average discharge of 4–5 m^3/s (**Figure 4**). The Powerhouse Spring discharge averaged about 0.25 m^3/s. After reservoir impounding, the estimated discharge of Big Spring increased from 10 to 16 m^3/s [26]. Few researches on the springs and reservoir water level fluctuations and sedimentology suggest that the springs' water passing under the foundation is independent of reservoir water elevation and depends on tail water elevation although some suggestions are against this conclusion [18].

The location of the two springs is aligned with a tectonic lineament trend SW-NE. Besides, the joint sets at the Big Spring location (**Figure 6**) shows the same general trend. This direction is parallel to the average direction of regional compressive stress in the Zagros Belt and indicates the general trend of extensional fractures forming normal to anticlinal structures

Figure 5. General layout of the dam site (left) and downstream view of Big Spring (right). The numbers in the left picture stand for: 1—Karun river, 2—reservoir, 3—Big Spring, and 4—cutoff wall.

Figure 6. A close view of Big Spring (left) and existing joint cavity (right).

pre or synchronous to folding [27]. The existing karst channel shows parallelism to the joint walls. The vertical dip of joint system on the right abutment could be another factor to support spring recharge through the reservoir area by very far upstream sources.

Another case of problem in the dam site is related to the construction of a new underground powerhouse regarding its time as the reservoir was impounded nearly 25 years before it. The newly designed structure (No. 2 powerhouse) was situated in the left bank of the dam and was excavated in the middle Asmari formation. Rock exposures displayed some signs of karstification such as open or filled cavities and small solution channels along the structural elements. Fortunately, no direct connection with tectonic elements was found in the left abutment. However, some inflow increase was observed in the newly excavated power house cavern [28], which lasted to the present.

As it was assumed basically that bedding planes were the major discontinuities of the site, the orientation of the No. 2 powerhouse was set perpendicular to the strike of bedding [20]. Fracture system in the left abutment was characterized with very wide to moderate spacing with the evidence of karstification along the joint planes. Some of the solution openings were filled with crystalline calcite and gypsum. All of this evidence supports increase of water inflow through the fractures system. The surface rock at the powerhouse slope belongs to the higher part of the Middle Asmari Formation. Based on the observations, joint sets were the main potential cause of rock instabilities of the cavern [28]. The measured permeability of rock strata was moderate to very high according to various available data with some signature of karstification. Although, Lugeon tests carried out in the geotechnical boreholes were limited to 4 LU, calculations showed a water inflow to the cavern up to 1500 l/min that necessitated creation of a gout curtain between cavern and reservoir. The last challenge related to the existing geologic structure was the potential for the mechanical initiation of a rock fall or slide during and after the dam construction. The construction phase is probably be one or two orders of magnitude higher than with natural causes [29], however, during the operation period, the phenomenon is possible as is the case for Karun-1 dam.

Figure 7. Rock sliding (in yellow) at right abutment of the dam.

Here, due to the downstream dip of bedding, which is unfavorable for a dam site, a natural potential for rock fall and slide on the dam abutments was predictable (**Figure 7**). This geological condition is of key importance since the southwestern (downstream side) limbs of the Zagros anticlines are usually very steeper than the northeastern limb due to the action of thrust faulting that affect the southwestern anticlinal limbs [30, 31]. New observations proved the subject and some rock slide and fall happened especially on the right abutment. There seems to be some flexuring along the bedding plane as well. The vertical tensional joints on the right abutment could amplify rock falling.

4. Karun-3 dam

Karun-3 is a 205 m high concrete dam constructed at 28 km to the south of Izeh City (**Figures 1** and **8**) in the southwestern Khuzestan Province. Electric power generation and flood control were the main objectives of the dam construction. Karun-3 power plant has a capacity of 2000 MW, with an average generation of 4137 GWh/y. This is a double curvature concrete arch dam, with a subsurface powerhouse located downstream the dam and power tunnels [32, 33]. It has symmetrical shape regarding the shape of its valley. The location and alignment of the dam are limited by geological and topographical features on both abutments (**Figure 8**). The dam reservoir is 60 km

Figure 8. Upstream view of Karun-3 dam site before (left) and after (right) construction.

long with storage volume of 3×10^9 m^3. The dam was impounded in November 2004. The presence of a hazardous slope named G2M above takeoff yard and a big downstream spring named as Abol-Ghasem Spring is of the main concern here.

The dam body was constructed on Asmari formation on the southwest limb of Keyf Malek Anticline that is surrounded by elevated anticlines including Lapeh (in the northeast) and Monghast (on the southwest) [32, 34]. The reservoir area is underlain by Pabdeh, Asmari, Gachsaran, and Agha-Jari formations. The Asmari formation limestone is the main water-bearing formation at the dam site and reservoir area with a potential for karst development similar to that discussed previously (part1) and is reported for other dam sites in the Zagros Fold Belt and in this chapter (Karun-1 and Marun dams). The Keyf Malek anticline, on which the dam lies, is mainly consisted of lower Asmari outcrops (**Figure 9**).

It is made up of interbedded limestone and marly limestone with porosity values between 1 and 15.7% that imply medium to extremely porous rocks. The limestone is generally light gray to light brownish gray, fine to medium grained, strong to very strong [32]. The bedding strikes NW-SE with low dip at the anticline crest. On the southwestern limb, the dip of layers is very steep up to 80° due SW (**Figure 10**). The northeastern limb of the anticline has a dip of about 70°–80°. The fold axis shows a slight plunge toward the southeast (141°/06°). Regular joint sets are developed and these have consistent orientations across the project area. A major NW-SE trending fault named as Doshab Lori Fault passes within 500 m to the southwest of the dam site. Another major thrust fault cuts the northeastern limb of the anticline creating an overturned syncline between the Keyf Malek and Lapeh anticlines (**Figures 9** and **10**). The major seismically active faults in the study area [35] are presented in **Figure 11**. Some small faults cut the anticline parallel and normal to its axial trend.

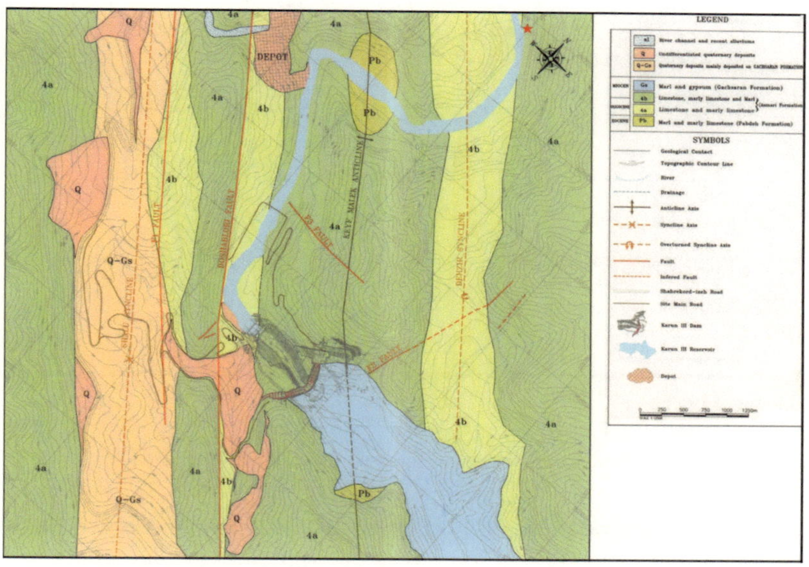

Figure 9. Geological map of the dam site area. Red star is the location of Abol-Ghasem Spring [36].

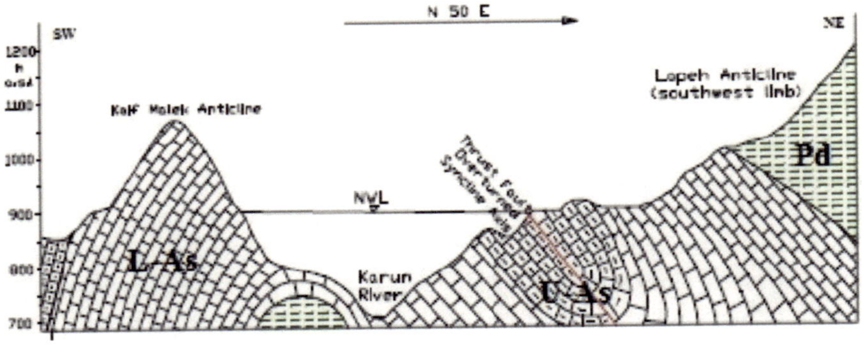

Figure 10. Geological section of Keyf Malek anticline [34].

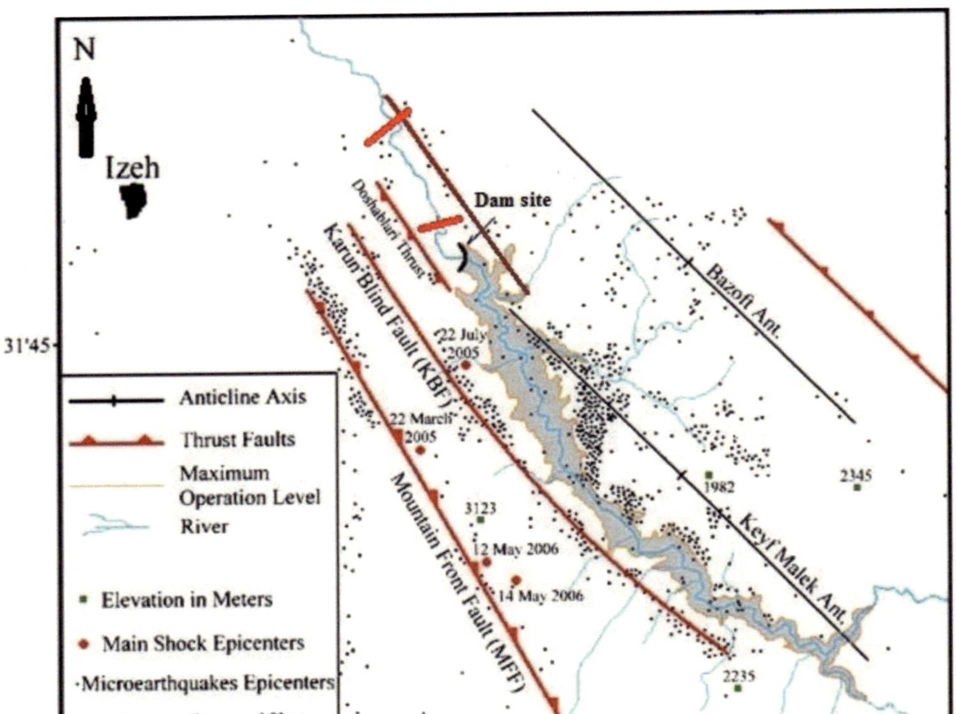

Figure 11. Regional faults and distribution of earthquake epicenters around Karun-3 dam [35].

The strike and dip of the strata vary only slightly over the dam and powerhouse sites. The bedding has a fairly flat inclination at the top of the fold, but become rapidly steeper to the west of the axis, where it generally dips 60°–85° southwest (**Figure 12**). Mapping indicates that several strongly developed joint sets characterize the bedrock in the area.

http://dx.doi.org/10.5772/intechopen.78742

Figure 12. Downstream view of Keyf Malek anticline (left) and longitudinal fault cutting its southwestern limb (right).

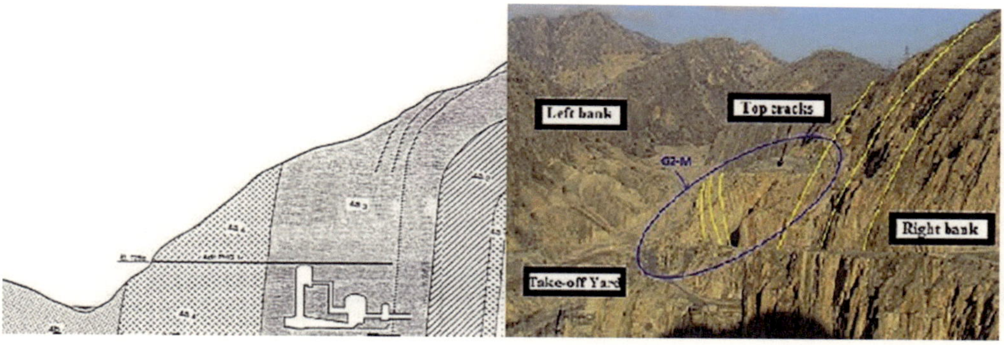

Figure 13. Geological section (left) and downstream view of G2M (right) rock mass.

As was mentioned above, instability of G2M above takeoff yard and leakage through downstream springs in particular Abol-Ghasem Spring are the main subjects of dam behavior in Karun-3 project. The rock slope, called G2M, is placed at the top of access road to the spillway on the right bank of the dam (**Figure 13**). The slope that is composed of limestone, marly limestone and marls besides developed tensional cracks created a potentially sliding mass.

Generally speaking, the slope was probably formed by displacement and collapse of the layers consequently, pushing front layers toward the river valley [33, 37]. Alternation of the competent (limestone) with incompetent (marl and marly limestone) layers at G2M slope, besides the presence of thrust faults on both flanks of Keyf Malek anticline is most probably the main reason for instability of the G2M. It is very probable that a hidden blind fault cuts the core of the anticline regarding very sudden change in dip of the bedding on the southwestern limb of

Keyf Malek anticline. This structural feature is very common in the Zagros folds specially in highly deformed and stressed regions such as Izeh fault zone in which the dam is located [36]. High fracturing of the rocks at the dam site is also a sign of the governing role of fault activity in the study area. Successive occurrence of thrust faults cutting the region and overturning of the Lapeh Anticline northeastwards is an indication of extensive tectonic deformation of the region resulted in high and dense fracturing of the Keyf Malek anticline. Additionally, the apparent break in the bedding integrity in the southwest limb of the anticline might be a sign of longitudinal faulting along its strike (**Figure 12**). It should be noted that Doshab Lory fault that runs between the Keyf Malek anticline and Mongasht anticline is a back thrust with a movement in the direction opposite to the general direction of regional tectonic transport. Since, such structures are generally an indication of fault propagation folds, intense fracturing between fore thrust and back thrust is reasonable, which is almost seen for the Karun-3 dam site. As mentioned before, Karun-3 dam site lies in the northern part of Izeh fault (shear) zone that is distinguished by a variety of thrust and dextral strike-slip faults [14] so that their interaction created a complex highly deformed and sheared geological region.

Another challenging subject in the dam region is the presence of Abol-Ghasem Spring downstream of the dam site. It is located at about 2.5 km downstream of the dam site, near the right bank of the Karun River (**Figure 9**). Its water was recharged by Lapeh anticline karstic aquifer, before impounding with varying seasonal discharge nearly between 0.5 and 1.5 m^3/s, and increased to about 2.5 m^3/s after reservoir impoundment. Additionally, another seasonal spring about 70 m downstream of the Abol-Ghasem Spring was changed to a permanent one after the dam impounding. It seems that the both limbs of Keyf Malek anticline are potential paths for seepage equally [34]. Karstification of the northeastern limb of the Keyf Malek anticline along with thrust faulting intensified the seepage at the spring. The presence of an overturned syncline between the two anticlines indicates intense tectonic compression in the region that could result in dense fracturing in the existing two anticlines. Transverse faults cut across the anticline that is evident through sharp and sudden change of Karun River course downstream the dam site might facilitate the occurrence of springs.

5. Marun dam

The Marun dam site is located on the Marun River in the Khuzestan Province approximately 19 km northeast of Behbahan City (**Figures 1** and **14**). The dam was commenced in 1997 and completed in 1999 with a height of 165 m, a crest length of 345 m and reservoir volume of about 1200 million cubic meters. As a rock fill dam, it is the second highest embankment dam in Iran. Its main purpose was flood control, water storage, and a total of 145 MW power generation. It also provides a dependable water resource for irrigation of 55,000 hects of downstream farm lands. The dam site is located on the limestone of the Asmari Formation, of Oligo-Miocene age. The formation is divided into the lower, middle, and upper Asmari formation with a total thickness of 370 m. It consisted of strong limestone partly interbedded with thin layers of shale. The whole formation at the dam site is characterized by karstification evidence.

The Marun dam was built on the northeastern flank of Khaviz anticline in the Zagros fold belt. The foundation of the dam consists of thick-bedded limestone of the Asmari formation

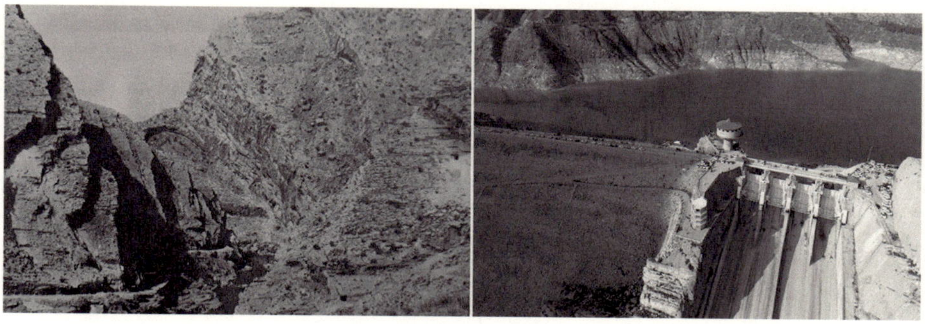

Figure 14. Upstream view of Marun dam site before (left) and after (right) construction.

with alternation of shale, marlstone gypsum, and anhydrite [38, 39]. The beds strike parallel to the dam axis trending NW-SE and average dip of approximately 35° due NE (**Figure 15**). The rock is regularly well jointed. Although fairly homogenous, the rock shows anisotropic permeability due to karstification of limestone. The rock strata at the site comprise a series of karstic limestones interbedded with water sensitive marls, which dip toward the reservoir. The main geological structures of the region include folds and faults aligned parallel to the main folding axis of NW-SE trend. The reservoir basin is centered mostly along the southwest flank of a broad northwest trending syncline. This feature forms a broad structural basin, approximately 9 km wide and 14 km long. At the dam site, two major joint sets are seen, the first parallel to the bedding, and the second perpendicular to it. However, a special set of

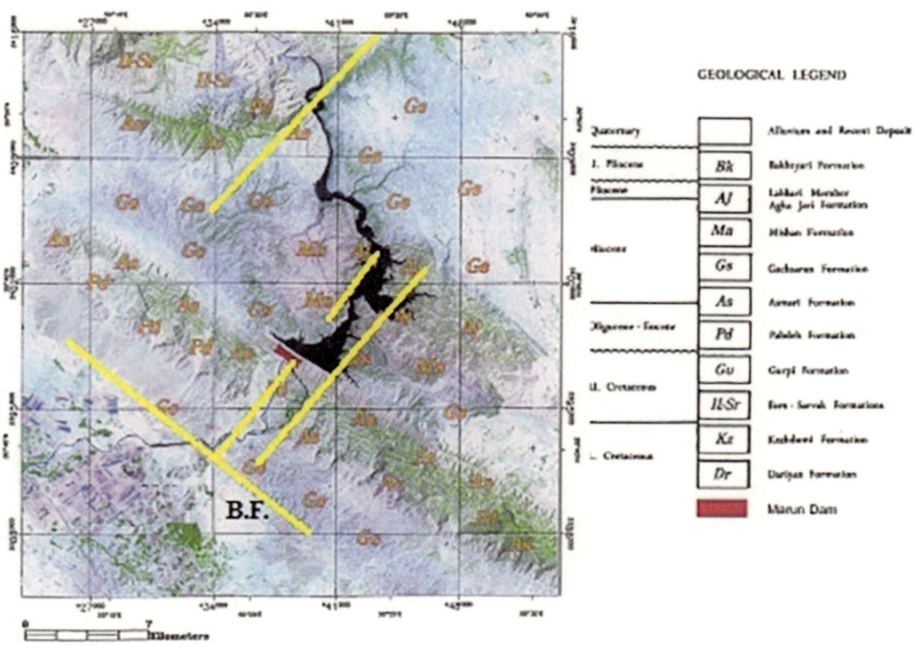

Figure 15. A general layout of Marun dam and appurtenant structures [39]. BF is Behbahan thrust fault.

fractures classified as fracture swarms [40] is also recognized here. Fracture swarms are nearly large-scale features, which dissect significant parts of the stratigraphic sequence. In the Khaviz anticline, fracture swarms are represented by faults with displacements of a few meters up to 150 m, and are associated with relative narrow damage zones with locally very high fracture frequency [40]. Three large thrust faults cut the region with a NW-SE general trend. These are named as Behbahan, Arajan, and Tashan faults [14]. A hidden fault cuts the core of Khaviz anticline parallel to Behbahan fault as well. While Karun-3 dam site lies in the northern part of Izeh fault zone, the Marun dam site is located at the southernmost part of it. The dam site is also affected by active faulting as indicated by its seismic activity especially for Tashan Fault.

The first and prominent problem encountered in Marun dam site was leakage through a diversion tunnel named as second diversion tunnel during the first impounding. Immediately after impoundment, considerable leakage was observed in the pressure tunnel (**Figure 16**) and efforts to open the stop logs failed. At the same time, an embankment was constructed and subsequent grouting controlled the leakage in the pressure tunnel. Old karst channels along a vuggy zone cut by the second diversion tunnel were reactivated and leakage occurred [39]. The total amount of water leakage through the left bank of Marun dam was about 10–15 m^3/s. The unlined second diversion tunnel had a key role in connecting reservoir with karst conduit system.

The embankment was overtopped with increasing water elevation, and considerable leakage of up to 7 m^3/s occurred from weak zones upstream of the plug. The major flow of approximately 4.5 m^3/s was from two large solution channels and leakage around the concrete plug.

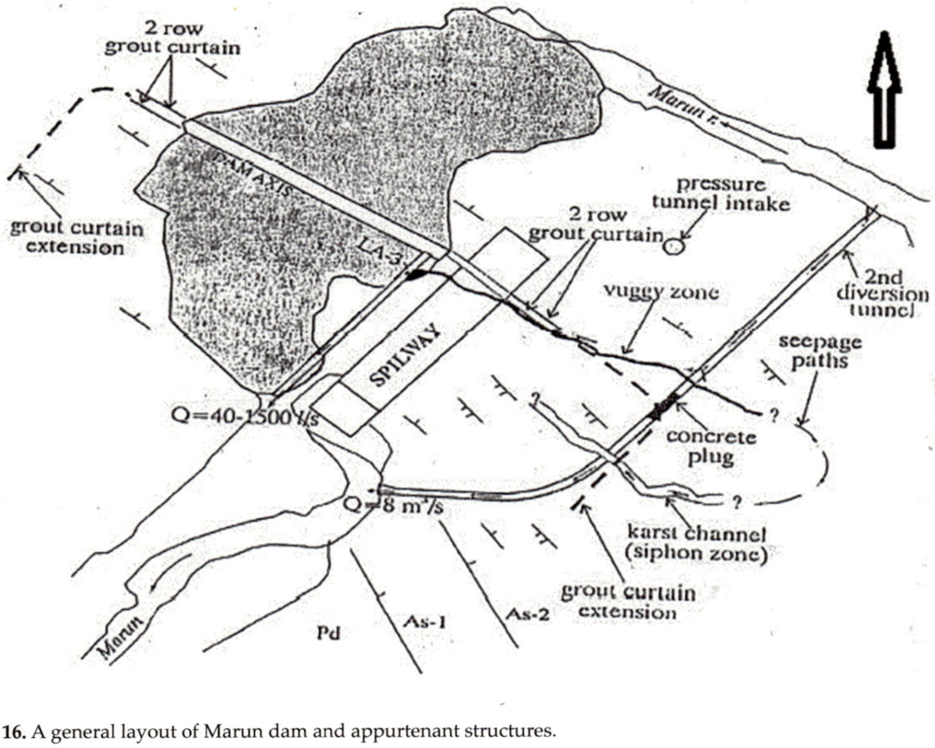

Figure 16. A general layout of Marun dam and appurtenant structures.

The remaining flow was from the access tunnel and the grouting adit (2.3 m^3/s). The total amount of water leakage through the left bank of Marun dam was about 10 m^3/s. The water entered the fracture system upstream of the plug and passed along the fissures, washed out the marls interbeds forming large cavities. These are master cross joints traversing the dam site rocks from upstream to downstream. Water was leaking into all tunnels and the dam, and all springs received their water from the same fissure as was reported [39]. In fact, these are fracture swarms as was defined above indicating the intense fracturing surrounding faults or narrow zones with a very high frequency of fractures. These major features are major conduits for fluid flow in the subsurface at some stage that is very favorable for karst development. This is almost similar to the karstic features mentioned for Karun-1 dam (abovementioned) although it is almost visible that karstification is developed more and complete at the Marun dam site. Here, the general trend of fracture warms is parallel to the general direction of the maximum compressive stress in the Zagros Fold Belt that produced tension-induced, open fractures resulting in significant secondary permeability to the rock. Intersection of longitudinal fractures (vuggy zone) in the Khaviz anticline with transverse fractures (fracture swarms) amplified permeability of the rock and simplified water leakage through the left abutment during the first reservoir impounding. It is very probable that seepage paths were formed along the fracture swarms parallel to the second diversion tunnel.

The second problem in Marun dam site was rock fall occurred along the left abutment recently [41]. Accordingly, stability studies and treatment are being conducted. The variation in shape of the rock blocks obtained from the joint data assessment indicates that the potential for dislodging rock blocks is medium to high. The horizontal joint sets in combination with the very steep slope face on the left bank result in blocks with a high rock fall hazard. The prevailing role of fracture system in rock slope instability at the dam site is again clear [41].

6. Gotvand dam

The Gotvand or Upper Gotvand dam as the highest rock fill dam in Iran is located in the north of Khuzestan Province (**Figures 1** and **17**). It was constructed across the Karun River in the north of Shushtar city. It is a 178 m high rock fill dam with central clay core and a crest length of 760 m. The normal reservoir level is situated at 232 m above sea level at which the reservoir reaches a capacity of 4.5 billion m^3 and around 90 km in length. The Gotvand dam has the hydropower plant with the highest energy production capacity in the country. The main purpose of the project is the production of (4250 GWh) hydroelectric energy. The impoundment of the reservoir started in July 2011 and by 2014, the reservoir water level had reached 223 m a.s.l. The main problems in the dam site and reservoir area are instability of the abutments [42] and with a minority, probable seepage potential through the foundation and abutments [43]. The dam foundation and part of the right abutment is underlain by Agha-Jari Formation of Mio-Pliocene Age (**Figures 17** and **18**). Agha-Jari lithology consists of gray, calcareous sandstone with gypsum veins and red marlstone and siltstone. Its rocks contain veins of gypsum usually associated with claystone beds. They are naturally soluble and can lead to excessive seepage. The left abutment of the dam is composed of Bakhtyari

Figure 17. An upstream (eastward) view of Gotvand dam site and the outcropped formations; Bakhtyari (Bk), Agha-Jari (Aj), and Gachsaran (Gs). DRM is displaced rock mass.

Conglomerates of Pliocene Age that displays almost horizontal bedding (**Figure 17**). The Bakhtyari Formation is wholly composed of terrigenous, clastic sediments ranging from silt to conglomeratic boulders. Fractures within this formation are usually vertical and have relatively large openings up to several meters. It is also composed of part of the right abutment as a dislocated and ruptured block (DRM). Along the northern margin of the river, Gachsaran layers are thrust over the Agha-Jari layers by Pir-Ahmad fault (**Figures 18** and **19**). The dam site area is dominated by an anticlinal structure comprising Kuh-e Reshteh and Kuh-e Charkhineha in the north and south of Karun River, respectively [44]. Both of

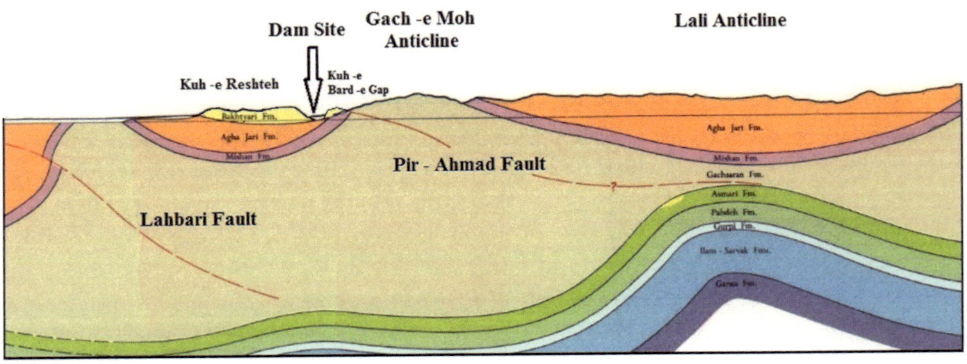

Figure 18. Geological section across the Gotvand dam region.

http://dx.doi.org/10.5772/intechopen.78742

Figure 19. A close view of Pir-Ahmad Fault in the northern side of Kuh-e Reshteh anticline.

them are composed of Bakhtyari formation. An elongated asymmetrical structure named Gach-e Mun or Gach-e Moh anticline occupies the north side of the dam site in a general NW-SE direction. Its south flank is obscured by the presence of Pir-Ahmad fault [44]. Kuh-e Reshteh is the southern part of the Bakhtyari outcrop on which the dam left abutment is placed. The Lahbari active fault passes along the contact between this outcrop and southern plain (**Figure 18**) and defines the mountain-plain boundary in the Dezful Embayment structural unit [14].

In fact, it forms the northern boundary of the Dezful Embayment (foreland basin of the Neogene molasse of the Agha-Jari-Bakhtyari formations). The river bed is occupied by a small tight anticline composed of Agha-Jari layers with an E-W trend. The anticlinal fold axis is parallel to a fault having the same E-W trend. Agha-Jari formation that composed the dam foundation contains three main joint sets. One set is parallel to the bedding plane. Severe change of joint dips is most probably due to tight folding of the layers and faulting along the river course. The joint surfaces are mostly smooth, polished, and slicken-sided with varying dip angles.

As mentioned above, one of the challenges at Gotvand dam is instability of the dam abutments [42] and its reservoir banks [44]. The dam abutments include considerable volumes of dislocated rock mass of Bakhtyari formation. This is more critical in the right abutment as the rock mass is highly disturbed and deformed (DRM) and apparently illustrates a rock topple (**Figures 17** and **20**). On the other hand, the rock mass in the left abutment shows evidence of a rock slide. The DRM area is specified by extensive development of joints and cracks and its

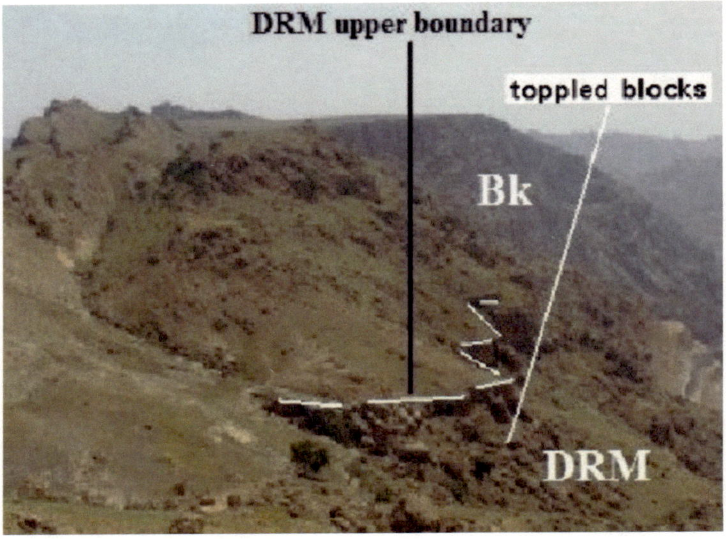

Figure 20. An upstream view of the displaced rock mass (DRM) on the right abutment [42].

extent along the dam axis was estimated between 150 and 200 m [45]. Consequently, water seepage was considered as an extra problem in the right abutment. Lugeon tests in this zone indicated values more than 60 due to the extent of fracturing. The prevailing hydromechanical behavior of the rocks at this abutment was dilatation and washout based on geotechnical investigations [44, 45].

The second problem expected for the Gotvand dam site, is leakage through its foundation via Agha-Jari layers [46]. These layers contain veins of gypsum usually associated with claystone beds. They appear as thin films on the beddings and along joint surfaces with a maximum thickness of 2 cm. The action of Lahbari and Pir-Ahmad thrust faults in the

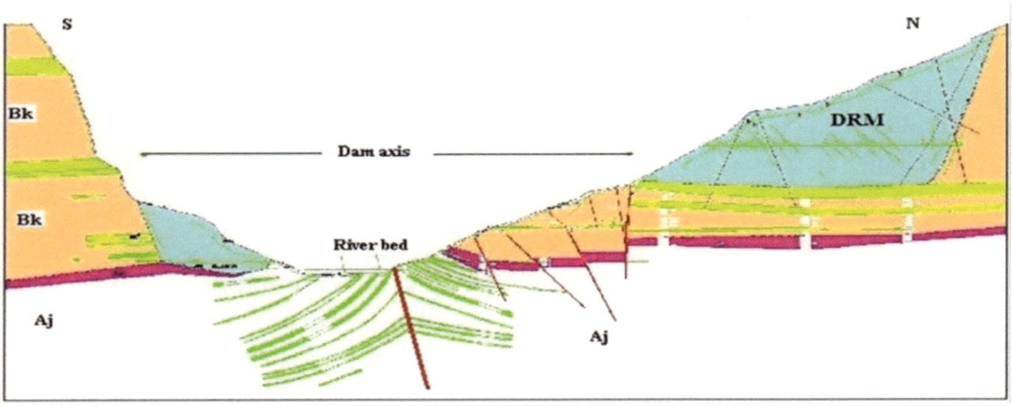

Figure 21. Simplified geological cross section along the dam axis [44].

south and north of the dam site, respectively, caused severe compression resulted in tight folding of the Agha-Jari layers (**Figure 21**). Subsequently, simultaneous compression of the competent sandstone layers and incompetent siltstone and mudstone layers in the Agha-Jari formation resulted in flexural-slip along the bedding planes, accompanied by shearing and jointing of rocks.

Continual development of forced-folding widens joint apertures that make them favorable for gypsum precipitation. Results of Lugeon test for Agha-Jari rocks at the dam foundation show an average value of 6 up to a maximum of 30 LU. Laboratory experiments estimated the solution rate for the core-drilled gypsum of the dam foundation as 2.49 cm/year. As the number of fractures in gypsum layers increases, solution processes progress. It is calculated that the dissolution of gypsum veins increase the mass equivalent permeability up to 75–300 times depending on aperture width and spacing. The availability of fractures and fissures are primary and main factors for solution progress. In this regard, gypsum veins observed in Agha-Jari formation at the dam foundation could be threats to safety and proper performance of the dam.

7. Conclusions

The knowledge of geological structures gives a reasonable in-sight in to dam construction studies. Site geology and availability of various geologic data obtained from site investigation is a key point in dam construction. It is also clear that the geological nature of different sites is not the same and depends on local and regional geology. Existing experience in dam construction projects shows that geological structure plays an important role in dam site geology and imposes major limitations on dam behavior during and after construction stages. In Karun-1 dam, for example, seepage through two springs downstream the right abutment are related most probably to existing shear zone at that abutment. Besides, rock falling and sliding on the dam abutments due to the downslope dip of bedding is another structure-related problem at the dam site. In Karun-3 dam, almost similar problems are observed but here, the presence of two sub-parallel thrust faults with opposite dip direction resulted in high stresses in the rock that intensified fracturing and subsequent permeability. Variability and complexity of geological structures regarding their tectonic situation result in different scenarios regarding dam's behavior. In this regard, the intersection of longitudinal and transverse faults in the vicinity of Marun dam caused the development of well and highly densed joint systems in the dam abutments that facilitated karstification. This in turn increased permeability of the rocks that was followed by extensive seepage during the dam construction. The role of geologic structure on geotechnical properties of dam sites is clearly seen in cases such as Gotvand Dam. Here, high fracturing and instability in the right abutment caused costly treatments including various stabilizing works such as grouting, geomembrane, and shotcrete. Finally, all the abovementioned examples indicate the impact of geological structure on various procedures either during constructing a dam or after its completion and close relation between structural geology and rock behavior.

Acknowledgements

The author would like to thank Mr. N. Bazgir, deputy of Dam Development and Operation and Mr. M.R. Shamsaie, the managing director of Khuzestan Water and Power Authority (KWPA) for their support and permission for publishing this article. Mrs. Roya Ghilav (KWPA) is also appreciated for reviewing the original manuscript.

Author details

Arash Barjasteh

Address all correspondence to: barjasteh@hotmail.com

Khuzestan Water and Power Authority (KWPA), Ahvaz, Iran

References

[1] Buljan R, Prelogovicz E. The significance of structural and geological relationship assessment in the construction of the Ombla underground hydroelectric power plant. Rudarsko-geološko-naftni zbornik. 1997;**9**:17-22. UDC:551.24.551.44:621.311:627.824 (497.13)

[2] Milanović PT. Water Resources Engineering in Karst. Boca Raton: CRC Press; 2004. 312p

[3] Cosgrove JW, Hudson JA. Structural Geology and Rock Engineering. London: Imperial College Press; 2016. 533p. ISBN 9781783269563

[4] Milanović PT. Dams and reservoirs in Karst. In: van Beynen PE, editor. Karst Management. Dordrecht: Springer; 2011. pp. 47-73. DOI: 10.1007/978-94-007-1207-2

[5] Oberlander T. The Zagros Streams: A New Interpretation of Transverse Drainage in an Orogenic Zone. New York: Syracuse University Press; 1965. 168p. ASIN: B0007DMUYE

[6] Summerfield M. Global Geomorphology. New York: Longman, Harlow; 1991. 537p. ISBN-13: 978-0582301566

[7] Oberlander TM. Origin of drainage transverse to structures in orogens. In: Morisawa M, Hack JT, editors. Tectonic Geomorphology. Boston: Allen and Unwin Publications; 1985. pp. 155-182

[8] Milanović PT. Optimal management of groundwater for meeting seasonally fluctuating demands. In: Parriaux A, editor. Proceedings of IAH Congress, Water Resources in Mountainous Regions. IAH Memoirs, Vol. XXII, Part 2, Lausanne, Switzerland, Ecole Polytechnique Federale de Lausanne; 1990. pp. 933-940

[9] McQuarrie N, Stock JM, Verdel C, Wernicke BP. Cenozoic evolution of Neotethys and implications for the causes of plate motions. Geophysical Research Letters. 2003;**30**(20): 20-36

[10] Stöcklin J. Structural history and tectonics of Iran. American Association of Petroleum Geologists Bulletin. 1968;**52**:1229-1258

[11] Alavi M. Tectonics of the Zagros orogenic belt of Iran: New data and interpretations. Tectonophysics. 1994;**229**:211-238

[12] Roustaei M, Nissen E, Abbassi M, Gholamzadeh A, Ghorashi M, Tatar M, et al. The 2006 March 25 Fin earthquakes (Iran)—Insights into the vertical extents of faulting in the Zagros Simply Folded Belt. Geophysical Journal International. 2010;**181**:1275-1291

[13] Bosold A, Schwarzhans W, Julapour A, Ashrafzadeh AR, Ehsani SM. The structural geology of the high central Zagros revisited (Iran). Petroleum Geoscience. 2005;**11**:225-238

[14] Berberian M. Master blind thrust faults hidden under the Zagros folds: Active basement tectonics and surface morphotectonics. Tectonophysics. 1995;**241**:193-224. DOI: 10.1016/s0014-5793(01)03293-8

[15] Ziaie Moayed R, Izadi E, Fazlavi M. In-situ stress measurements by hydraulic fracturing method at Gotvand dam site, Iran. Turkish Journal of Engineering and Environmental Sciences. 2012;**36**:179-194

[16] Harza Engineering Company. Karun River Development Project. Feasibility Report, I, II, III. Chicago; 1967

[17] Lahmeyer International and Moshanir Consulting Company. Geological and Rock Mechanics Report. Shahid Abbaspour Dam, Powerhouse No. 2 Project; 1994. 76pp

[18] Ghobadi MH, Khanlari G, Djalali H. Seepage problems in the right abutment of the Shahid Abbaspour dam, southern Iran. Journal of Engineering Geology. 2005;**82**:119-126

[19] Mahab-Ghods Consulting Engineers Company. Detailed safety Evaluation Report of Karun Arch Dam. Khuzestan Water and Power Authority; 1985. 65pp. unpublished report

[20] Lahmeyer International and Moshanir Consulting Company. Hydrogeological report. Vol. 1/2., Shahid Abbaspour Dam, Powerhouse No.2 Project. Khuzestan Water and Power Authority. Addendum No. 7. 81pp

[21] Aliashgari HA, Berry DR. Rock fall analysis of the Shahid Abbaspour Dam left abutment and stability approaches. In: Proceedings of the 3rd Iranian Rock Mechanics Conference (3IRMC); Tehran, Iran: Amir Kabir University of Technoloy; 16-18 October 2007. pp. 866-870

[22] Zistab Consulting Engineers Company. Engineering Geology Report. Shahid Abbaspour Dam, Phase 1 Studies for Increasing Operating Level. Khuzestan Water and Power Authority; 2014. 78p

[23] Barjasteh A. Hydrogeomechanical Impacts of Increasing the Operating Level of Karun-1 (Shahid Abbaspoor) Dam, Iran. Abstracts of 35th IGC, Cape Town, South Africa. 29 August to 2 September 2016. 2p

[24] Zare M. A study on the strong motions recorded in the Masjed-Soleyman dam site during the Andeka Earthquake 2002, MW5.6. In: CD of proceedings of 73rd ICOLD annual meeting; Tehran, Iran. 1-6 May 2005. 11p

[25] Ebrahimi M, Tatar M. Induced seismicity around Masjed Soleyman Dam, South West of Iran. In: CD of proceedings of 15th World Conference on Earthquake Engineering (15WCEE); Lisbon, Portugal: Sociedade Portuguesa de Engenharia Sismica (SPES); 24-28 September 2012. 10p

[26] Milanović P. Geological Engineering in Karst. Belgrade: Zebra Publishing; 2000. 347p

[27] McQuilan H. Fracture pattern on Kuh-e Asmari Anticline, Southwest Iran. The American Association of Petroleum Geologists Bulletin. 1974;**58**(2):236-246

[28] Lahmeyer International. Report No. 2 Shahid Abbaspour Dam, Powerhouse No. 2 Project; 1997. 72pp

[29] Hoek E, Kaiser PK, Bawden WF. Support of Underground Excavations in Hard Rock. Netherlands: Balkema Publishers; 1995. 232 p. ISBN 90 5410 186 5

[30] Barjasteh A. Geologic structure control on seepage potential of the Upper Gotvand Dam, Iran. CD of proceedings, 81st ICOLD annual meeting. Seattle, USA; 2013. 9p

[31] Barjasteh A. Active faulting at the Gotvand Dam based on seismic and geotechnical data. In: Proceedings of 7th International Conference on Seismology and Earthquake Engineering (SEE7); Tehran, Iran: International Institute of Earthquake Engineering and Seismolgy (IIEES); 18-21 May 2015. 8p

[32] Koleini M. Engineering geological assessment and rock mass characterization of the Asmari formation (Zagros range) as large dam foundation rocks in southwestern Iran [Ph.D thesis]. South Africa: University of Pretoria; 2012. 243p

[33] Gharouni-Nik M. Rock slope stability analysis for a slope in the vicinity of take-off yard of Karun-3 Dam. In: Chen et al., editors. Landslides and Engineered Slopes–From the Past to the Future. Vol. 2. Proceedings of the 10th International Symposium on landslides and engineered slopes; 30 June–4 July 2008; Xi'an, China. London: Taylor & Francis Group; 2008. pp. 1725-1729

[34] Damough NA, Zarei H. On the geological problems of Karun-3 dam reservoir regarding to seepage subject. ICOLD Annual meeting, International Symposium on Dams and Sustainable Water Resources Development. Hanoi, Vietnam; 2010. 17p

[35] Kangi A, Heidari N. Reservoir-induced seismicity in Karun-3 dam (Southwestern Iran). Journal of Seismology. 2008;**12**:519-527. DOI: 10.1007/s10950-008-9104-4

[36] Sadd Tunnel Pars Consulting Engineers Company. Geology and Assessment of Instabilities in Dam and Power House Area. Safety Studies of Karun-3 Dam and Power House Area. Khuzestan Water and Power Authority; 2015. 87p

[37] Ghazvinian AH, Moradian ZA. A case study of failure mechanism and support measures of slopes against sliding and toppling failure. In: van Cotthem A, Charlier R, Thimus JF, Tshibangu JP, editors. Multiphysics Coupling and Long Term Behavior in Rock Mechanics. Proceedings of the International Symposium on ISRM. Eurock 2006; Liege, Belgium: Balkema Publication; 9-12 May 2006. pp. 309-314

[38] Zarei HR. Principal engineering geological problems in Marun Dam project, Iran. In: CD of proceedings of the 76th ICOLD Annual Meeting; Sofia, Bulgaria. June 2-6 2008. 10p

[39] Ghobadi MH, Karami R. Solubility of limestone and seepage problems in the left abutment of the Marun dam, southwest of Iran. Journal of Engineering Geology. 2009; **3**(1):615-632

[40] Wennberg OP, Svånå T, Azizzadeh M, Aqrawi AMM, Brockbank P, Lyslo KB, et al. Fracture intensity vs. mechanical stratigraphy in platform top carbonates: The aquitanian of the Asmari formation, Khaviz anticline, Zagros, SW Iran. Petroleum Geoscience. 2006;**12**:235-245

[41] Koleini M, Van Rooy JL. Falling rock hazard index: A case study from the Marun Dam and power plant, south-western Iran. Bulletin of Engineering Geology and the Environment. 2011;**70**:279-290. DOI: 10.1007/s10064-010-0327-6

[42] Sharifi BM, Salehi D, Hazrati M, Mahjoub D. Dynamic stability analysis of the Upper Gotvand dam right abutment against earthquake (in Persian). In: CD of proceedings of the 1st International and 3rd National Conference on Dam and Hydropower Plants, IWPCO; Tehran, Iran. 8-9 Feb 2012. 9p

[43] Sharifzadeh M, Nateghi R, Kiyani M. Hydro-mechanical modeling of seepage in Gotvand dam foundation. In: Shao JF, Burlion N, editors. Thermo-Hydromechanical and Chemical Coupling in Geomaterials and Applications. Proceedings of the 3rd International symposium GeoProc. Hoboken: Wiley; 2008. DOI: 10.1002/9781118623565.ch73

[44] Barjasteh A. The impact of active faulting on the geotechnical properties of the Upper Gotvand dam, southwest Iran. Bulletin of Engineering, Geology and Environment. 2017:1-14. DOI: 10/1007/s10064-017-1163-8

[45] Yaghoobi RJ. Applying real time grouting control method in sedimentary rock with Gotvand Dam [unpublished M.Sc. Thesis]. Royal Institute of Technology (KTH); 2010. 107p

[46] Sadrekarimi J, Kiyani M, Fakhri B, Vahdatifard MJ, Barari A. Seepage analysis of upper Gotvand dam concerning gypsum karstification (2D and 3D approaches). Frontiers of Architecture and Civil Engineering in China. Higher Education Press and Springer. 2010;**5**(1):71-78. DOI: 10.1007/s11709-010-0083-5